Handbook of Materials Circular Economy

Seeram Ramakrishna · Brindha Ramasubramanian

Handbook of Materials Circular Economy

Springer

Seeram Ramakrishna
Department of Mechanical Engineering, Center for Nanotechnology and Sustainability, College of Design and Engineering
National University of Singapore
Singapore, Singapore

Brindha Ramasubramanian
Department of Mechanical Engineering, Center for Nanotechnology and Sustainability, College of Design and Engineering
National University of Singapore
Singapore, Singapore

ISBN 978-981-97-0588-7 ISBN 978-981-97-0589-4 (eBook)
https://doi.org/10.1007/978-981-97-0589-4

This Springer imprint is published by the registered company Springer Nature Singapore Pte Ltd.
The registered company address is: 152 Beach Road, #21-01/04 Gateway East, Singapore 189721, Singapore

Contents

Chapter 1
Introduction to Materials Circular Economy

Abstract This chapter introduces the concept of "Materials Circular Economy" (MCE) and explores its scope, benefits, and classification of materials. The chapter discusses the parameters used to evaluate sustainable materials, including carbon footprint, global warming potential, energy consumption, resource efficiency, and waste reduction potential. Additionally, the role of engineered materials and biomaterials in achieving circularity is examined. The chapter also presents a comprehensive overview of the steps required to foster a circular economy in consumer products. Furthermore, the global scenario and the impact of digital technologies on the implementation of circular economy principles are discussed.

Keywords Circular design strategies · Material life cycle analysis · Digitalization in circular economy · Circular economy initiatives · Resource efficiency and carbon footprint

1.1 Introduction

The depletion of global resources as a result of human activity is raising worry among businesses and governments all over the world. Our economies' production and consumption systems are harming the planet's natural system, putting additional strain on the planet's resources. Manufacturing the billions of things for modern living requires large amounts of raw materials and energy, and the trash created in the process is discharged into the air, water, and land, causing harm to key ecosystems [1].

In a world where every individual consumes resources and generates waste at the same rate, our planet would struggle to meet the soaring demand for resources [2]. To address this problem, we must transform from linear to circular economies, where resources and products are kept at the highest possible value and lifespan, to achieve more with less. As practical examples of circular economies emerge, we are gaining a clearer understanding of what this transition will entail. The primary goal is to extend the lifespan of resources and products, keeping them functional for as long

S. Ramakrishna and B. Ramasubramanian, *Handbook of Materials Circular Economy*,
https://doi.org/10.1007/978-981-97-0589-4_1

as possible, thereby reducing the amount of waste generated. This shift is crucial for the survival of our planet and its ecosystems [3].

1.1.1 Basics of "Materials Circular Economy"

A linear materials management system is a "take-make-dispose" approach in which raw materials are removed, processed, and changed into products that are then discarded as trash. This method generates a considerable quantity of garbage and depletes natural resources. A closed-loop system, on the other hand, seeks to preserve the value of resources by limiting waste, reusing goods, and recycling materials to generate new products. In traditional textile business, cotton often becomes waste. Yet, in the fashion industry's closed-loop system, materials are recycled and items reused, cutting waste and demand for new goods. [4].

In simple terms, Circular Economy (CE) is an economic model that prioritizes the continual use and regeneration of resources to minimize waste and maintain the sustainable environment. This model has been illustrated in the Fig. 1.1a. An application of the circular economy principles specifically to materials and their management, focused on creating closed loops and minimizing waste throughout the materials' life cycle is called Materials Circular Economy (MCE). The concept of circular economy has been shown in Fig. 1.1b.

Circular economy principles are the set of guidelines that aim to achieve the goals of a circular economy. Here are five principles that underlie a circular economy:

1. Products should prioritize extended lifecycle, focusing on reuse, repair, and recycling to reduce waste

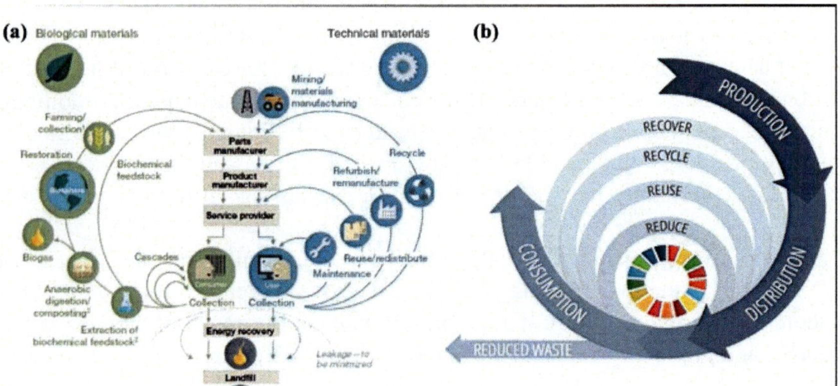

Fig. 1.1 **a** Circular economy business model [8]. **b** Concept of circular economy [9]

2. Minimization of resource usage through material substitution and improved energy efficiency [5].

3. Both producers and consumers should be held responsible for the full product lifecycle, promoting circular design and responsible resource management.

4. Stakeholder collaboration to share resources, utilization of digital technologies to drive innovation, and implementation of a transparent supply chain are essential.

5. Regulations should be implemented to support circular practices, including suitable policies and eco-design standards[6, 7].

1.1.2 Scope and Benefits of Circular Economy

By reducing the burden of material and energy consumption on our planet, circular economy can help protect ecological goods and services from the pollution and waste generated by consumer lifestyles. Moreover, it can help us reduce the per capita resource consumption, ensuring that there are enough resources for everyone's well-being. Many developing countries still require resources to grow and prosper, and the circular economy can help secure these resources. However, realizing a circular economy requires the engagement of society, along with significant invention and innovation, the creation of new businesses, technologies, and governance systems [10]. Such transformative changes offer immense potential to stimulate employment and increase the demand for skilled workers, generating value for society. Nevertheless, this transition demands new ways of thinking, social systems, engagement strategies, and institutions, calling for an evolved society [11]. The scope of the circular economy encompasses a range of industries and sectors, from manufacturing to construction, agriculture to retail. In essence, any sector that consumes resources and produces waste can benefit from circular economy principles [12].

1.1.3 Classification of Materials for MCE

Materials can be classified for Materials Circular Economy (MCE) based on their circularity potential, which refers to the ability of a material to be reused, repaired, and recycled within a closed-loop system [13, 14].

Linear Materials

Linear materials are single-use materials that cannot be reused, mended, or recycled. In a linear economy, they have poor circularity potential and can lead to waste and resource depletion. The scientific importance of linear materials stems from their influence on the environment and human health over their entire life cycle, from extraction to disposal. Linear materials, such as disposable packaging or single-use items, are frequently designed to give convenience or efficiency in the short term. Its long-term impact on the environment and society, however, may be enormous, since they contribute to natural resource depletion, pollution, and greenhouse gas emissions. The utilization of linear materials carries significant consequences. Firstly, such materials are frequently non-renewable and limited, signifying that their utilization and extraction can result in the depletion of natural resources. Secondly, the discarding of linear materials as waste can generate considerable environmental impacts, such as contributing to landfills and marine pollution. Lastly, the manufacturing and transportation of linear materials can trigger greenhouse gas emissions, thus aggravating climate change [15]. Current data indicates a rapid escalation in the production and consumption of linear materials. For example, global plastic production has skyrocketed from 1.5 million tonnes in 1950 to over 359 million tonnes in 2018. Similarly, global aluminum production has ascended from 15 million tonnes in 1970 to over 63 million tonnes in 2018. These trends are anticipated to persist, with a projected upsurge in demand for linear materials in tandem with population and economic growth [15].

Despite advancements in the recyclability of plastics, some materials still pose a challenge or cannot be recycled at all. Polystyrene foam, commonly known as Styrofoam, is a non-recyclable material due to its low-density nature, making it difficult to sort during the recycling process. Shocking statistics reveal that in the United States, less than 10% of polystyrene foam is recycled. Thin plastic films, frequently used in food packaging, are another example of non-recyclable materials as they can tangle in machinery or contaminate other recyclables. Currently, globally, less than 5% of plastic films are recycled [16].

Mixed-material packaging such as juice boxes or snack pouches, containing plastic, metal, and paper, poses another challenge in recycling due to its complexity. Currently, only a few specialized recycling programs accept these materials, resulting in less than 5% of mixed-material packaging being recycled globally. Ceramics and glassware are also non-recyclable materials as they have a different melting point than glass bottles, which can damage the recycling equipment. Finally, materials contaminated with hazardous or toxic substances like chemicals or food waste are not recyclable and pose a risk to workers in recycling facilities. Figure 1.2a depicts linear material flow.

Semi-circular Materials

Semi-circular materials are recyclable or reusable, but they necessitate further refining or treatment to deliver optimal recycling or reusing. These materials have a modest degree of circularity potential. Some types of plastics and paper are examples of semi-circular materials. PVC is a thermoplastic synthetic polymer that is commonly used in building, power lines, and vinyl flooring. It includes chlorine atoms, which complicate the recyclability. A specific procedure known as "thermal

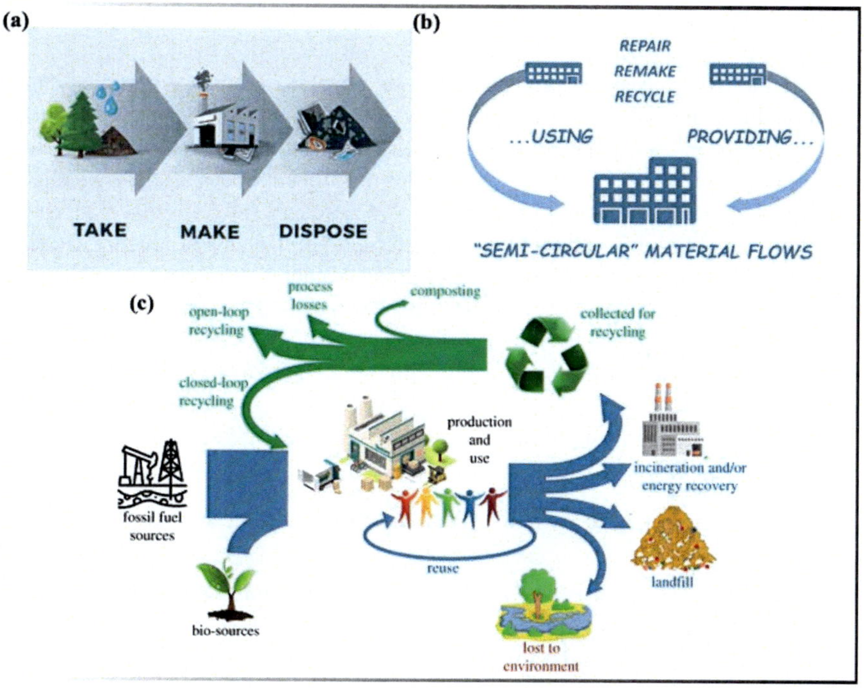

Fig. 1.2 **a** Material flow in a linear pattern [22]. **b** Material flow in a semi-circular pattern [23]. **c** Material flow in a circular pattern [24]

depolymerization" is required to recycle PVC. PVC is heated in this procedure to break it down into its basic constituents such as hydrogen, carbon, and chlorine. The resultant gases are separated and purified before the carbon and hydrogen are employed to manufacture new synthetic polymers. PS is a malleable plastic that is commonly used in packaging, disposable cups, and insulating materials [17]. Yet, because to its low density and enormous volume, recycling with typical methods is problematic. Generally, pyrolysis is applied to recycle PS. In the absence of oxygen, the plastic is heated, which is then condensed and purified to form new plastic materials.

As per a report published by the Ellen MacArthur Foundation, the global recycling rate of plastic packaging stands at roughly 30%, while 8% of it is being incinerated, and the remaining 62% is either disposed of in landfills or ends up polluting the environment. The report further indicates that paper and cardboard have a recycling rate of approximately 58% globally, with the remaining 42% either being incinerated or sent to landfills [18, 19]. Figure 1.2b illustrates semi-circular material flow.

Circular Materials

Fully circular materials are materials that can be recycled or reused without any loss in their quality or performance. These materials are considered to be the most sustainable due to their high level of circularity potential and minimal environmental impact. Glass, aluminium, and steel are a few examples of fully circular materials. Mechanical recycling is the most common method of processing fully circular materials, which involves shredding and melting the material to create new products [20]. For instance, crushed and melted glass can be utilized to manufacture new glass bottles, while melted aluminium can be utilized to make new cans. Furthermore, biodegradable materials are also considered to be fully cyclable [21]. Figure 1.2c demonstrates material flow in circular economy. Table 1.1 shows the type of material, and the key assessment criteria.

1.1.4 Parameters to Evaluate Sustainable Materials

When evaluating the circularity of a material, it is important to take into account a comprehensive set of eight parameters, including carbon footprint, global warming potential, energy consumption, resource efficiency, waste reduction potential, reusability, recyclability, and other environmental impacts [25].

Carbon Footprint

Carbon footprint is a measure of the total greenhouse gas (GHG) emissions caused by an individual, organization, event, or product, expressed in terms of carbon dioxide equivalent (CO_2e). It is a useful tool to evaluate the environmental impact of an activity or product, and to identify opportunities for reducing emissions and improving sustainability [26]. The formula to calculate the carbon footprint involves multiplying the amount of each GHG emitted by the relevant global warming potential factor and adding them up [27]. For example, to calculate the carbon footprint of a car, we would need to know the amount of fuel consumed and the emission factors for carbon dioxide, methane, and nitrous oxide. The formula would be (Table 1.2):

$$\begin{aligned}\text{Carbon footprint} = {} & (\text{fuel consumption}) \times (\text{carbon dioxide emission factor}) \\ & + (\text{fuel consumption}) \times (\text{methane emission factor}) \\ & \times (\text{global warming potential factor}) + (\text{fuel consumption}) \\ & \times (\text{nitrous oxide emission factor}) \times (\text{global warming potential factor})\end{aligned}$$

Table 1.1 Types of material, its preparation and key assessment parameters for circularity

Material type	Preparation method	Global warming potential (kg CO_2-e/kg material)	Carbon footprint (kg CO_2-e/kg material)	Emission factor (kg CO_2-e/kg material)
Linear	Extraction or production of virgin materials	High	High	High
Example: petroleum-based plastics	Crude oil is extracted and refined to produce plastic pellets	1.3–3.7	6.1–20.1	1.3–3.7
Semi-circular	Recyclable or reusable with additional processing	Moderate	Moderate	Moderate
Example: some types of plastic and paper	Plastics may require sorting and cleaning before recycling. Paper may require de-inking or treatment to remove contaminants	0.3–1.5	1.2–5.5	0.3–1.5
Circular	Recyclable or biodegradable with minimal processing	Low	Low	Low
Example: organic materials such as food waste, wood, and some types of bioplastics	Biodegradable materials can be composted, while recyclable materials can be easily melted or reshaped	0.1–0.3	0.3–0.8	0.1–0.3

Table 1.2 List of major activities in a manufacturing industry and their formula

Activity/industry	Formula
Energy use	Carbon footprint = Energy consumption × Emission factor
Transportation	Carbon footprint = Distance travelled × Vehicle fuel efficiency × Emission factor
Food	Carbon footprint = Food consumption × Emission factor
Buildings	Carbon footprint = Building energy consumption × Emission factor
Products	Carbon footprint = (Raw material extraction + Manufacturing + Transportation + End-of-life disposal) × Emission factor

Solved Problems

X1.1 A car consumes 10 L of gasoline per 100 km, and the emission factors for carbon dioxide, methane, and nitrous oxide are 2.3, 0.01, and 0.004 g per liter of fuel, respectively. The global warming potential factors for methane and nitrous oxide are 25 and 298 times that of carbon dioxide, respectively. Calculate the carbon footprint of the car per km.

Solution: Carbon footprint = (10 L) × (2.3 g/l) + (10 L) × (0.01 g/l) × (25) + (10 L) × (0.004 g/l) × (298)

= 230 g/km + 2.5 g/km + 11.9 g/km.

= 244.4 g/km.
So, the carbon footprint of the car would be 244.4 g of CO_2e per km travelled.

1.2 ABC Food Processing Plant processes various food products, including vegetables, fruits, meat, and dairy products. The plant uses electricity, natural gas, and diesel as its energy sources and generates waste from the production process. The plant also transports raw materials and finished products using trucks. Calculate the carbon footprint of the plant and identify reduction opportunities.

Given data

Electricity consumption: 10,000 kWh

Natural gas consumption: 500 GJ

Diesel consumption: 100,000 L

Distance travelled by trucks: 10,000 km

Waste generated: 10 tons.

Assumptions

The emission factors for electricity, natural gas, and diesel are 0.6 kg CO_2e/kWh, 56.1 kg CO_2e/GJ, and 2.68 kg CO_2e/liter, respectively.

The emission factor for truck transportation is 0.2 kg CO_2e/ton-km.

The waste disposal method is landfill, with an emission factor of 1.1 kg CO_2e/ton.

Solution

Step 1: Identify the emissions sources:

Electricity consumption: 10,000 kWh × 0.6 kg CO2e/kWh = 6000 kg CO_2e.

Natural gas consumption: 500 GJ × 56.1 kg CO2e/GJ = 28,050 kg CO_2e.

Diesel consumption: 100,000 L × 2.68 kg CO_2e/liter = 268,000 kg CO_2e.

Truck transportation: 10,000 km × 0.2 kg CO_2e/ton-km × 10 tons = 20,000 kg CO_2e.

Waste disposal: 10 tons × 1.1 kg CO_2e/ton = 11 kg CO_2e.

Step 2: Calculate the total carbon footprint:

Total carbon footprint = 6000 kg CO_2e + 28,050 kg CO_2e + 268,000 kg CO_2e + 20,000 kg CO_2e + 11 kg CO_2e = 322,061 kg CO_2e.

Step 3: Identify reduction opportunities:

Natural gas and diesel consumption carbon footprints are high compared to other emission sources. Therefore, the plant can optimize its energy use by upgrading its equipment and implementing energy-saving measures, such as insulation and lighting upgrades. Installation of solar panels to generate renewable energy on-site.

Emission factor and where to find it

An emission factor is a standardized metric that links the amount of pollutant discharged into the atmosphere with a particular activity or process. Essentially, it quantifies the volume of greenhouse gas emissions associated with a given unit of activity or process. The purpose of emission factors is to estimate the amount of greenhouse gas emissions that arise from various sources, including industrial operations, transportation, and energy use. Emission factors are typically expressed in units of mass per unit of activity, such as kilograms of carbon dioxide equivalent (CO_2e) per kilowatt-hour (kWh) of electricity generated or kilograms of CO_2e per liter of gasoline consumed. The determination of emission factors entails measurements and calculations that take into account a variety of factors, including the type of fuel used, the effectiveness of the equipment or process, and the characteristics of the emission source. Emission factors are sourced from a range of outlets, such as government agencies, industry associations, and academic research [28].

To locate an emission factor for a specific activity or process, you can refer to several databases or sources, including the Emission Inventory Improvement Program (EIIP) database of the U.S. Environmental Protection Agency (EPA) or the greenhouse gas inventory guidelines of the Intergovernmental Panel on Climate Change (IPCC). It's crucial to note that emission factors may differ based on the specific conditions and assumptions employed in their calculation [29]. Figure 1.3 presents the calculated statistics of carbon emissions by various global companies.

Note

The U.S. Environmental Protection Agency's Emission Inventory Improvement Program (EIIP) database: https://www.epa.gov/air-emissions-inventories/emission-inventory-improvement-program-eiip

The Intergovernmental Panel on Climate Change (IPCC) guidelines for greenhouse gas inventories: https://www.ipcc-nggip.iges.or.jp/public/2006gl/

The European Environment Agency (EEA) Emission Database for Global Atmospheric Research (EDGAR): https://edgar.jrc.ec.europa.eu/

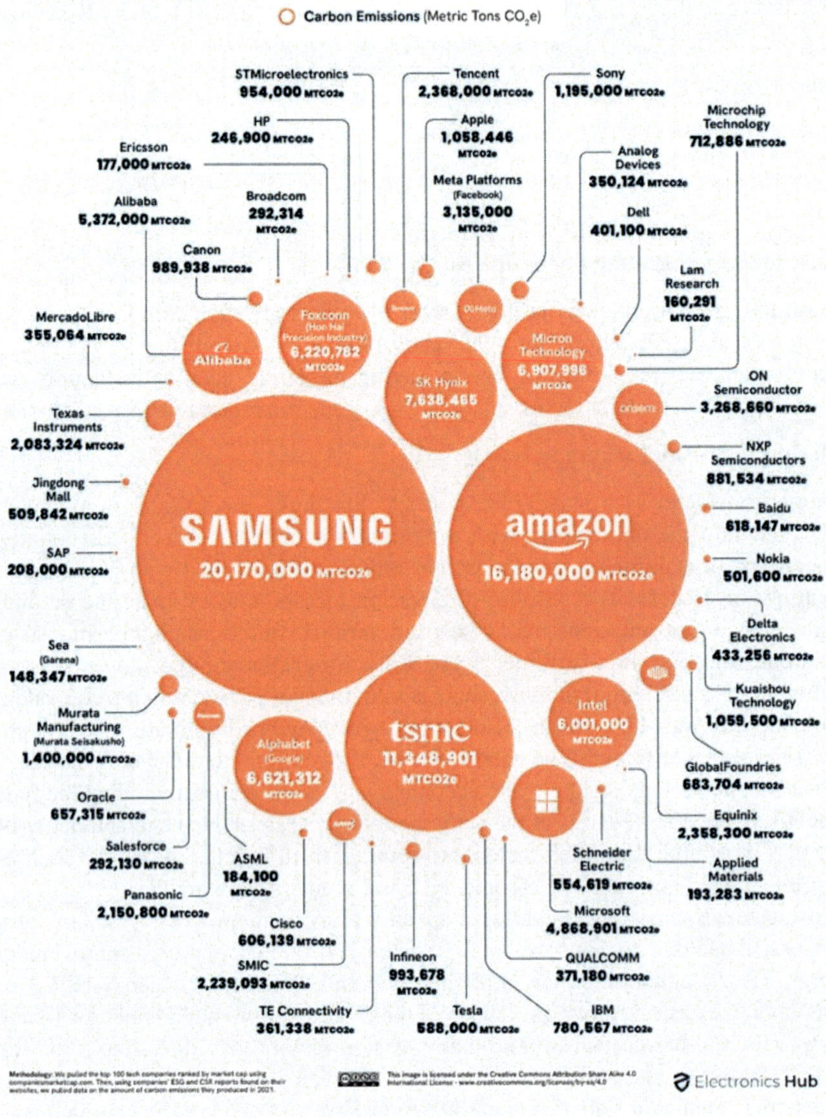

Fig. 1.3 Carbon emissions of Big Tech companies [30]

Global Warming Potential

The Global Warming Potential (GWP) is a measure of the environmental impact of greenhouse gas (GHG) emissions, notably in terms of global warming. The Global Warming Potential (GWP) allows scientists and policymakers to assess the warming effects of various GHGs based on their potential to trap heat in the atmosphere over a specified time period [31]. The GWP of a gas is calculated by comparing its warming potential to that of carbon dioxide (CO_2), which is assigned a GWP of 1. The formulas for calculating GWP depend on the time frame considered, with commonly used time frames including 20, 100, and 500 years [32]. The general formula for calculating GWP over a period of t years is:

$$\mathrm{GWPt} = (\mathrm{A1} \times \mathrm{GWP1} + \mathrm{A2} \times \mathrm{GWP2} + \cdots + \mathrm{An} \times \mathrm{GWPn})/\mathrm{ACO_2}$$

where A1 to An are the amounts of each gas emitted, GWP1 to GWPn are the GWPs of each gas over t years, and ACO2 is the amount of CO_2 emitted over t years. Some of the commonly used databases to fetch GWP include the Intergovernmental Panel on Climate Change (IPCC) Fifth Assessment Report, Ecoinvent, and the United States Environmental Protection Agency (EPA) greenhouse gas reporting program.

Radiative forcing (RF) is a measure of a greenhouse gas's (Table 1.3) or other climatic factor's ability to affect the Earth's energy balance, causing climate change. Watts per square meter (W/m^2) is the unit of measurement [33].

Problems

1.3 The concentration of a greenhouse gas in the atmosphere is given by the function

$$C(t) = 3te^{(-0.1t)}$$

Table 1.3 List of greenhouse gases and their corresponding global warming potential and atmospheric lifetime

Greenhouse gas	Atmospheric lifetime (years)	GWP (100-year time horizon)
Carbon dioxide (CO_2)	50–200	1
Methane (CH_4)	12	28–36
Nitrous oxide (N_2O)	121	265
Chlorofluorocarbons (CFCs)	Up to 100 years	4660–10,720
Hydrofluorocarbons (HFCs)	Up to 260 years	Up to 12,400
Perfluorocarbons (PFCs)	Up to 50,000 years	7390–12,200
Sulfur hexafluoride (SF6)	3200	22,800
Water vapor (H_2O)	9	0–0.04
Carbon monoxide (CO)	1–2	1
Nitrogen oxide (NO_x)	114	298
Ammonia (NH3)	7	90
Tropospheric ozone (O3)	Hours to weeks	22–56
Hydrogen (H_2)	0.1	0
Carbonyl sulfide (COS)	120	69
Fluorinated ethers (FEs)	Up to 100 years	Up to 14,800

where t is the time in years. Determine the GWP of this greenhouse gas over a 100-year time horizon, assuming that its radiative forcing capacity is given by:

$$\mathrm{RF(t)} = 0.02\mathrm{C(t)}$$

Solution

To calculate the GWP, we need to first determine the total radiative forcing over a 100-year time horizon, which is given by:

$$Rf_{total} = \int_{0}^{100} Rf(t)dt$$

Substituting the expression for RF(t), we get

$$Rf_{total} = \int_{0}^{100} 0.02C(t)dt$$

$$Rf_{total} = 0.02 \int_{0}^{100} 3te^\wedge(-0.1t)dt$$

$$Rf_{total} = 0.02\{30 - 300e^{-10}\}$$

$$Rf_{total} = 0.6 - 6e^{-10}$$

Next, to determine the GWP, which is defined as the ratio of the total radiative forcing of a greenhouse gas to that of carbon dioxide over a 100-year time horizon. The GWP of carbon dioxide is 1. Therefore:

$$\mathrm{GWP} = \mathrm{RF\_total/RF\_CO_2}$$

Substituting the GWP and radiative forcing values, we get

$$GWP = (0.6 - 6e^{-10})/1$$

Finally, to determine the rate of change of GWP with respect to time, we can differentiate the expression for GWP:

$$\frac{dGWP}{dt} = \frac{6e^{-10}}{10^{-10}}$$

$$\frac{dGWP}{dt} = 6 \times 10^{-20}$$

Therefore, the GWP of the greenhouse gas over a 100-year time horizon is 0.6–6e$^{(-10)}$, and its rate of change is $6 \times 10^{(-20)}$ per year.

Energy Consumption

Energy consumption refers to the amount of energy used to power homes, businesses, and industries. It is typically measured in units of kilowatt-hours (kWh) or joules (J). The formula to calculate energy consumption is:

$$\text{Energy Consumption} = \text{Power (in Watts)} \times \text{Time (in hours)}$$

$$\text{Energy Consumption} = \text{Current (in Amperes)} \times \text{Voltage (in Volts)} \times \text{Time (in hours)}$$

The above formulas can be used for both direct current (DC) and alternating current (AC) circuits. The resulting value will be in units of watt-hours (Wh) or kilowatt-hours (kWh) [34].

Embodied energy refers to the total amount of energy required to extract, manufacture, transport, and dispose of a material or product. It includes all the energy inputs needed to produce the material or product, such as energy used for mining, refining, processing, and transportation of raw materials, as well as energy used in manufacturing, packaging, and shipping.

Low embodied energy materials: These are materials that require minimal energy inputs for their production, such as wood, bamboo, natural fiber composites. These materials are suitable for engineering applications where minimizing environmental impact is a priority.

Medium embodied energy materials: These are materials that require moderate energy inputs for their production, such as glass, some types of metals, and certain types of plastics. These materials can be used in engineering applications where a balance between environmental impact and performance is required.

High embodied energy materials: These are materials that require significant energy inputs for their production, such as aluminum, steel, and concrete. These materials should be used sparingly in engineering applications where minimizing environmental impact is a priority [35].

Case study

Measuring the energy consumption of a company involves monitoring the amount of energy used by various equipment and processes in the facility (Table 1.4). Walmart, a well-known multinational corporation, has implemented an effective energy consumption monitoring program through its Sustainability 360 initiative, which was launched in 2005 with a focus on reducing the company's energy consumption and greenhouse gas emissions. To begin, Walmart identified electricity as the primary energy source in its stores, followed by natural gas for heating and cooling. Subsequently, the company installed energy meters in all of its stores, allowing for the measurement of energy usage across different systems and equipment in the facilities. The energy meters were able to accurately measure various aspects of energy usage such as flow rate, voltage, current, and power factor. The collected energy data was analyzed through specialized software tools to identify energy consumption patterns and opportunities for energy savings. The data showed that energy consumption was highest during peak business hours and on certain days of the week. Walmart benchmarked its energy consumption with other similar retailers to identify areas of inefficiency. The benchmarking revealed that Walmart's energy consumption was higher than the industry average. To further optimize energy usage, Walmart conducted energy audits to identify specific opportunities for energy savings. The energy audits identified several opportunities, such as implementing more efficient lighting systems, upgrading HVAC systems with more efficient equipment and controls, and optimizing store layouts to reduce energy waste [36].

Table 1.4 Different circularity considerations and their formula

Formula	Description	Example
Energy efficiency ratio	Calculates the ratio of output or benefit produced by a system or process to the amount of energy input required to achieve that output or benefit	Output/Energy input
Water use efficiency	Calculates the ratio of yield or output produced by an agricultural system to the amount of water input required to achieve that yield or output	Yield/Water input
Material efficiency	Calculates the ratio of output or benefit produced by a manufacturing process or product to the amount of material input required to achieve that output or benefit	Output/Material input
Carbon efficiency	Calculates the amount of carbon emissions produced per unit of output or benefit	Carbon emissions/Output
Labor productivity	Calculates the amount of output or benefit produced per unit of labor input	Output/Labor input
Land use efficiency	Calculates the amount of yield or output produced per unit of land used	Yield/Land use
Water productivity	Calculates the amount of yield or output produced per unit of water used	Yield/Water use
Nutrient use efficiency	Calculates the amount of nutrient uptake or use by a crop or agricultural system per unit of nutrient input	Nutrient uptake/Nutrient input
Time efficiency	Calculates the amount of output or benefit produced per unit of time input	Output/Time input
Financial efficiency	Calculates the amount of output or benefit produced per unit of financial investment or cost	Output/Financial investment

As a result of these measures, Walmart was able to achieve significant energy savings by reducing its energy consumption by 28% and saving over $200 million in energy costs. Furthermore, Walmart's energy consumption monitoring program helped the company identify and address inefficient equipment and processes, resulting in a more sustainable and cost-effective operation. This case study has been publicly reported by Walmart through its sustainability reports and other corporate communications, highlighting the success of its energy consumption monitoring program [37].

The following are some of the common steps and tools used to measure energy consumption, identified after reviewing 120 case studies:

1 Identify the energy sources: Determine the energy sources used in the company, such as electricity, natural gas, or diesel. This will help to identify the relevant measuring units and equipment.
2 Install energy meters: Install energy meters at strategic points in the facility to monitor the energy consumption of different systems and equipment. Energy meters can measure the flow rate, voltage, current, and power factor of the energy being used.
3 Collect and analyze data: Collect the energy consumption data from the energy meters and analyze it using software tools to identify trends, patterns, and opportunities for energy savings.
4 Benchmark energy consumption: Compare the energy consumption of the company with similar facilities in the same industry to identify areas of inefficiency.
5 Conduct energy audits: Conduct energy audits to identify specific opportunities for energy savings, such as upgrading equipment, improving insulation, or implementing energy-efficient practices [38].

Resource Efficiency

Resource efficiency is a measure of how efficiently resources such as energy, water, and materials are used to produce goods and services. It is an important concept for sustainable development, as it helps to minimize waste and reduce the environmental impact of economic activities. Resource efficiency can be calculated in a number of ways, depending on the specific resource being measured and the context of the calculation. In general, resource efficiency is measured by comparing the output or benefit of a process or activity to the amount of resources used to achieve that output or benefit. This can be expressed as a ratio, such as the amount of energy used per unit of output, or as a percentage, such as the percentage of raw materials that are recycled or reused in a production process [39]. There are many parameters that influence resource efficiency, including the type and quality of resources used, the efficiency of production processes, and the level of demand for the goods and services produced [40].

Waste Reduction Potential

The waste reduction potential is the amount of waste that can be reduced or eliminated by implementing various waste reduction strategies such as recycling, source reduction, composting, and reuse, (Table 1.5). Calculating this potential requires consideration of several technical terms and matrices, including the waste generation rate, waste composition, waste diversion rate, material recovery rate, and source reduction potential [41]. By using the formula,

$$\begin{aligned}\text{Waste Reduction Potential} = {} & \text{Waste Generation Rate} \\ & \times (1 - \text{Waste Diversion Rate}) \\ & \times \text{Material Recovery Rate}\end{aligned}$$

It is possible to determine the amount of waste that could potentially be reduced through these strategies. Other factors such as carbon footprint, energy consumption, water use, and life cycle assessment can also be considered to identify opportunities for waste reduction and resource efficiency in different stages of the product life cycle.

Problem 1.4

Company A generates 10 tons of waste per year, and its waste diversion rate is 50%, with a material recovery rate of 75%, the waste reduction potential would be:

$$\text{Waste Reduction Potential} = 10\ \text{tons/year} \times (1 - 0.5) \times 0.75$$
$$\text{Waste Reduction Potential} = 2.5\ \text{tons/year}$$

This means that by implementing waste reduction strategies, such as source reduction, recycling, and composting, the company could potentially reduce its waste generation by 2.5 tons per year.

1.1.5 Engineered Materials and Biomaterials

Engineered materials, also known as advanced or smart materials, are intentionally designed, and optimized to possess specific properties and functionalities. These materials are created through advanced manufacturing techniques, often involving the combination of different components or the alteration of microstructures at the atomic or molecular level. The primary objective is to achieve enhanced performance, durability, and functionality beyond that of traditional materials. Common types of engineered materials include composites, ceramics, metals, and polymers, each exhibiting unique characteristics such as high strength, improved electrical conductivity, exceptional heat resistance, enhanced chemical stability, and tailored optical properties [42].

Table 1.5 Material type, circularity criterias and real time application

Material type	Material name	Circularity index	Statistical availability	Promotion of material reuse and recycling	Real-time examples of implementation
Metals	Aluminium	High	Abundant	High	Aluminium cans recycled into new cans or other aluminium products, recycled aluminium used in construction and transportation industries
Metals	Copper	High	Moderate	High	Copper scrap recycled into new copper products, copper wire and tubing recycled into new wire and tubing products
Metals	Manganese	Low	Limited	Low	Manganese used in batteries, but not widely recycled due to technical limitations
Metals	Iron	High	Abundant	High	Scrap iron recycled into new iron and steel products, recycled iron used in construction and transportation industries
Metals	Nickel	Low	Moderate	Low	Nickel used in batteries and alloys, but not widely recycled due to technical limitations
Metals	Zinc	Low	Abundant	Low	Zinc used in galvanizing and alloys, but not widely recycled due to technical limitations
Metals	Lead	Low	Moderate	Low	Lead used in batteries and alloys, but not widely recycled due to environmental concerns

(continued)

Table 1.5 (continued)

Material type	Material name	Circularity index	Statistical availability	Promotion of material reuse and recycling	Real-time examples of implementation
Metals	Titanium	Low	Limited	Low	Titanium used in aerospace and medical industries, but not widely recycled due to technical limitations
Metals	Magnesium	Low	Limited	Low	Magnesium used in aerospace and automotive industries, but not widely recycled due to technical limitations
Metals	Gold	Low	Limited	Low	Gold used in electronics and jewelry, but not widely recycled due to low availability and high cost
Metals	Silver	Low	Abundant	Low	Silver used in electronics and jewelry, but not widely recycled due to low availability and high cost
Polymers	Polyethylene (PE)	Moderate	Abundant	Moderate	Plastic bags recycled into new bags or other plastic products, recycled PE used in building and construction materials
Polymers	Polystyrene (PS)	Low	Abundant	Low	Polystyrene foam used in insulation, but not widely recycled due to technical limitations
Polymers	Polyvinyl chloride (PVC)	Low	Abundant	Low	PVC used in pipes and other construction materials, but not widely recycled due to technical limitations

(continued)

Table 1.5 (continued)

Material type	Material name	Circularity index	Statistical availability	Promotion of material reuse and recycling	Real-time examples of implementation
Polymers	Polypropylene (PP)	Moderate	Abundant	Moderate	Plastic containers recycled into new containers or other plastic products, recycled PP used in furniture and household items
Polymers	Polyethylene terephthalate (PET)	High	Abundant	High	PET bottles recycled into new bottles or other plastic products, recycled PET used in textiles and carpeting
Polymers	Nylon	Low	Moderate	Low	Nylon used in textiles and automotive parts, but not widely recycled due to technical limitations
Polymers	Polycarbonate (PC)	Low	Limited	Low	PC used in electronics and automotive parts, but not widely recycled due to technical limitations
Polymers	Acrylonitrile butadiene styrene (ABS)	Low	Limited	Low	ABS used in electronics and automotive parts, but not widely recycled due to technical limitations
Glass	Soda-lime glass	High	Abundant	High	Glass bottles and jars recycled into new glass containers, glass cullet used in concrete and asphalt production
Glass	Borosilicate glass	Low	Limited	Low	Borosilicate glass used in laboratory equipment and cookware, but not widely recycled due to technical limitations

(continued)

Table 1.5 (continued)

Material type	Material name	Circularity index	Statistical availability	Promotion of material reuse and recycling	Real-time examples of implementation
Ceramics	Porcelain	Low	Limited	Low	Porcelain used in household items and construction materials, but not widely recycled due to technical limitations
Ceramics	Stoneware	Low	Limited	Low	Stoneware used in household items and construction materials, but not widely recycled due to technical limitations
Ceramics	Earthenware	Low	Limited	Low	Earthenware used in pottery and construction materials, but not widely recycled due to technical limitations
Composites	Carbon fiber reinforced polymer (CFRP)	Low	Limited	Low	CFRP used in aerospace and automotive industries, but not widely recycled due to technical limitations
Composites	Glass fiber reinforced polymer (GFRP)	Low	Limited	Low	GFRP used in construction and automotive industries, but not widely recycled due to technical limitations
Composites	Natural fiber reinforced polymer (NFRP)	Low	Limited	Low	NFRP used in automotive and construction industries, but not widely recycled due to technical limitations

(continued)

Table 1.5 (continued)

Material type	Material name	Circularity index	Statistical availability	Promotion of material reuse and recycling	Real-time examples of implementation
Composites	Metal matrix composite (MMC)	Low	Limited	Low	MMC used in aerospace and automotive industries, but not widely recycled due to technical limitations
Composites	Ceramic matrix composite (CMC)	Low	Limited	Low	CMC used in aerospace and industrial applications, but not widely recycled due to technical limitations

The process of engineering these materials involves various steps. Firstly, engineers carefully select the base materials, such as polymers, metals, ceramics, or composites, based on the desired properties and application requirements. Different materials offer specific advantages and can be combined to create hybrid materials with enhanced properties. Advanced manufacturing techniques, such as casting, forging, extrusion, or additive manufacturing (e.g., 3D printing), are then employed to process and shape the materials into the desired form, achieving specific geometries and microstructures. Engineers also have control over the microstructure of engineered materials, enabling manipulation at the atomic or molecular level. This control includes aspects like grain size, phase composition, crystal orientation, or the introduction of nanostructures to enhance mechanical, electrical, or thermal performance. Furthermore, surface modifications, such as coatings, plating, or surface patterning, are applied to improve functionality or interaction with the environment, thereby enhancing properties like wear resistance, corrosion resistance, or biocompatibility. Finally, engineered materials undergo rigorous testing and validation, evaluating mechanical properties, thermal properties, chemical stability, and other relevant parameters to ensure they meet the desired specifications and exhibit reliable performance [43].

Biomaterials are designed for interaction with biological systems in medical and healthcare applications. They exhibit biocompatibility, derived from natural or synthetic sources, and possess tailored properties such as mechanical strength, degradation rate, and surface characteristics. Biodegradable polymers, ceramics, metals, hydrogels, and bioactive glasses are common biomaterial types used in implants like artificial joints, cardiovascular stents, and tissue scaffolds. Achieving biocompatibility involves selecting non-toxic materials, while surface modifications enhance interactions through coatings, bioactive molecules, and specific topographies. Mechanical properties match target tissues, degradation rates are controlled via composition or additives, and fabrication techniques like 3D printing enable precise control over structure and properties [44]. Different materials have different end of life considerations (Table 1.6).

1.1.6 Steps to Increase Circular Economy in Consumer Products

The process involves several important steps to ensure safe and circular material choices in design processes:

Classifying materials based on the post-use phase: This step involves evaluating whether materials are suitable for a biological or technical cycle. Materials suitable for a biological cycle can return to the environment, while materials suitable for a technical cycle can be reused, transformed, or recycled after use.

Seeking information on chemical composition: Engaging with suppliers to obtain detailed information about the chemical composition of materials is important for understanding their potential environmental and health impacts. This information helps assess the safety and suitability of materials in the context of circularity.

Conducting materials screening: Screening materials for known hazards using tools like MaterialWise allows for the identification of specific chemical substances of concern. This screening helps in making informed decisions regarding material selection, substitution, and design optimization [45].

Prioritizing recycled or responsibly sourced materials: Giving preference to recycled materials or those sourced from responsible suppliers can reduce the environmental impact associated with raw material extraction. This step encourages the use of materials derived from waste streams or properly managed renewable resources.

Considering circular design principles: Circular design principles involve assessing material combinations, durability, repairability, and end-of-life considerations. By designing products with these principles in mind, materials can be utilized efficiently, extending their lifespan and facilitating their recovery or recycling.

Planning for proper treatment or recovery in the after-use phase: This step focuses on designing products to enable material recovery, disassembly, biodegradation, or suitable treatment after their use phase. Considering the end-of-life scenarios helps ensure that materials can be effectively managed, reducing waste and enabling their reintroduction into the economy [46].

Figure 1.4 illustrates the implementation Strategic measures for Circular Economy in Singapore.

Table 1.6 Types of materials and end of life considerations

Material type	Material selection considerations	End-of-life considerations	World initiatives	Company initiatives
Metals	• Strength and durability	• Recycling and reuse	• United Nations Sustainable Development Goals (SDGs)	• ArcelorMittal: Aims to produce 30% of steel from recycled materials by 2030
	• Corrosion resistance	• Proper disposal options	• Circular economy initiatives	• Novelis: Pioneered a closed-loop recycling system to achieve 80% recycled content in its products
	• Conductivity	• Recovery of valuable metals	• Paris Agreement	• Nucor Corporation: Implements electric arc furnace (EAF) technology that allows the recycling of steel scrap
Polymers	• Mechanical properties	• Recycling and incineration	• The Ellen MacArthur Foundation's New Plastics Economy	• Dow Chemical: Developing innovative recycling technologies to advance the circularity of plastics
	• Chemical resistance	• Landfill diversion	• Plastic waste management initiatives	• Procter & Gamble: Committed to using 100% recyclable or reusable packaging by 2025

(continued)

Table 1.6 (continued)

Material type	Material selection considerations	End-of-life considerations	World initiatives	Company initiatives
	• Cost effectiveness	• Reduction of microplastics	• Ocean cleanup projects	• Loop Industries: Specializes in chemical recycling of PET plastics, enabling high-quality recycling
Ceramics	• High-temperature stability	• Recycling and reuse	• UNESCO World Heritage initiatives	• Corning Incorporated: Developed ceramic waste management systems to recycle glass and ceramic materials
	• Hardness and wear resistance	• Proper disposal options	• Conservation of archaeological ceramics	• RAK Ceramics: Utilizes advanced waste treatment facilities to reduce environmental impact
	• Chemical inertness	• Reduction of hazardous waste	• Sustainable Ceramics initiatives	• Porcelanosa Group: Emphasizes eco-design and sustainable production processes to minimize waste
Composites	• Specific mechanical properties	• Separation of components	• Global composite recycling initiatives	• Boeing: Focuses on recycling carbon fiber composites from end-of-life aircraft to reduce waste

(continued)

Table 1.6 (continued)

Material type	Material selection considerations	End-of-life considerations	World initiatives	Company initiatives
	• Lightweight	• Proper disposal options	• Circular economy initiatives	• Toray Industries: Develops carbon fiber recycling technologies and works towards zero waste in manufacturing
	• Corrosion resistance	• Composite material recycling technologies	• Sustainable composites initiatives	• Gurit Holding AG: Implements closed-loop recycling systems for composite materials
Biomaterials	• Biocompatibility	• Biodegradability	• United Nations Sustainable Development Goals (SDGs)	• NatureWorks LLC: Produces biopolymers from renewable resources, reducing dependence on fossil fuels
	• Tissue regeneration properties	• Composting	• Bioeconomy initiatives	• Medtronic: Focuses on developing bioabsorbable medical devices that eliminate the need for removal surgeries
	• Low toxicity	• Reduction of medical waste	• Sustainable agriculture initiatives	• BASF: Works on developing sustainable biomaterials for various applications, such as biodegradable packaging

Fig. 1.4 Circular economy initiatives implemented by Singapore [47]

1.1.7 World Scenario and Play of Digital Technologies

The Ellen MacArthur Foundation estimates that the global circular economy market could grow to $4.5 trillion by 2030. Despite projections that the global demand for materials will double by 2060, the circular economy approach has the potential to reduce global material demand by 32% by 2050. By embracing circular principles, the European Union could create 700,000 new jobs by 2030. Additionally, the circular economy model could reduce global greenhouse gas emissions by 39% by 2050. In the United States, the recycling and reuse industry provides employment to over 600,000 individuals and generates an impressive $36.6 billion in annual wages. The global scenario for promoting a circular economy is marked by various initiatives and collaborations. The European Union leads the way with its Circular Economy Action Plan, encompassing waste reduction targets, resource efficiency measures, and promotion of eco-design and recycling. The Ellen MacArthur Foundation, a global leader in circular economy advocacy, drives change through research, education programs, and the development of circular economy standards. The United Nations acknowledges the significance of circular economy principles in achieving sustainable development, as reflected in SDG 12 on responsible consumption and production [48]. Many countries have formulated national strategies, such as the Netherlands' comprehensive approach that emphasizes closing material loops, promoting circular business models, and fostering resource management innovation. Collaborative initiatives and partnerships like the CEC LAC and CE100 network facilitate knowledge sharing and collective action. Additionally, circular economy hubs and innovation centers serve as platforms for collaboration, research, and the development of circular business models. These diverse initiatives collectively contribute to the global recognition of the importance of transitioning to a circular economy, paving the way for a more sustainable and resource-efficient future. Transitioning towards a digitalization based circular economy in Indonesia has been illustrated in Fig. 1.5a.

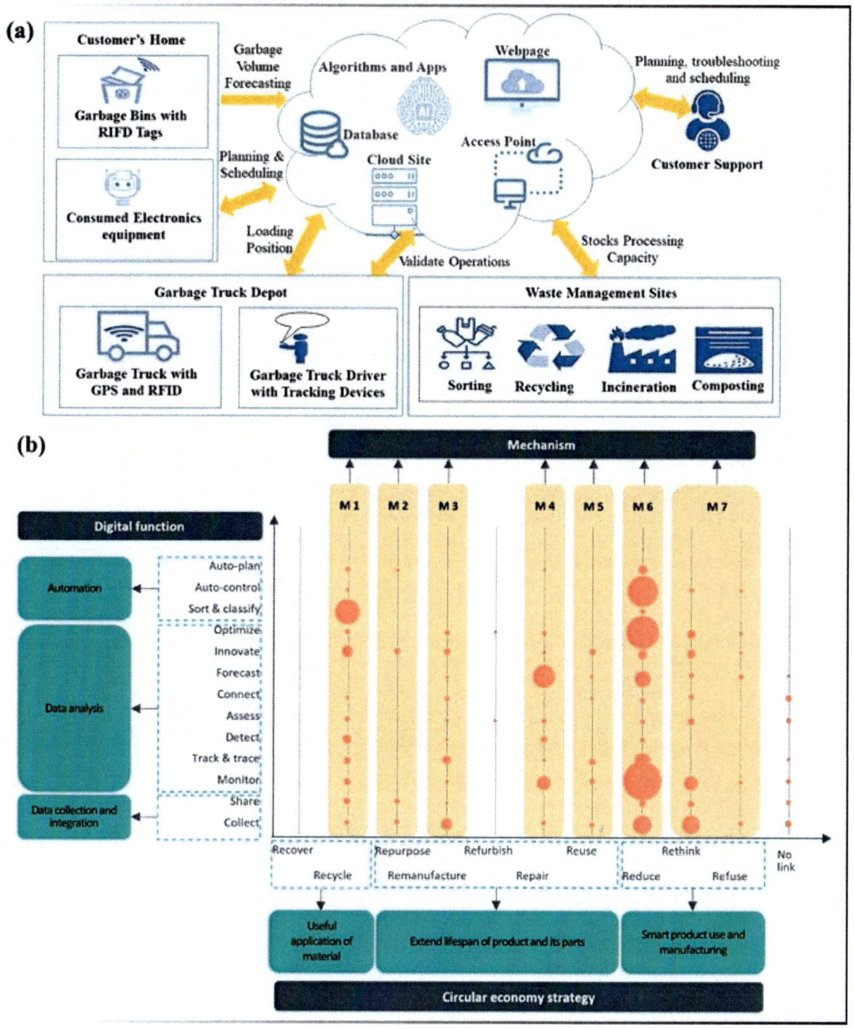

Fig. 1.5 **a** Digitalized circular economy [50]. **b** Framework of digital technologies for the circular economy: enabling digital functions and mechanisms [51]

Digital technologies play a vital role in promoting the circular economy through resource efficiency, supply chain optimization, and circular product design. Examples of technologies that aid in the circular economy include IoT sensors for monitoring product usage, blockchain for supply chain transparency, AI for supply chain optimization, 3D printing for on-demand production, cloud-based platforms for efficient collaboration, digital twin technology for optimization, Augmented Reality (AR) for visualization, and Product Lifecycle Management (PLM) software for product data management. These technologies enable businesses to reduce waste, improve resource management, and promote circularity in their operations [49]. Figure 1.5b presents an illustration of the framework of digital technologies for the circular economy, encompassing digital functions and mechanisms.

References

1. S. K. Ghosh, in *Circular Economy: Global Perspective*, ed. S. K. Ghosh, Springer, Singapore, 2020, pp. 1–23.
2. Ellen MacArthur Foundation, in *An Introduction to Circular Economy*, eds. L. Liu and S. Ramakrishna, Springer, Singapore, 2021, pp. 397–417.
3. G. Gaustad, M. Krystofik, M. Bustamante and K. Badami, *Resources, Conservation and Recycling*, 2018, **135**, 24–33.
4. A. Stefanakis and I. Nikolaou, *Circular Economy and Sustainability: Volume 1: Management and Policy*, Elsevier, 2021.
5. M. Esposito, T. Tse and K. Soufani, *California Management Review*, 2018, **60**, 000812561876469.
6. M. F. Aguiar and D. Jugend, *Journal of Cleaner Production*, 2022, **365**, 132732.
7. W. R. Stahel, *The Circular Economy: A User's Guide*, Routledge, 2019.
8. Sustainable Materials Management – A Precursor to a Circular Economy | GreenBlue, https://greenblue.org/sustainable-materials-management-a-precursor-to-a-circular-economy/, (accessed July 22, 2023).
9. bkbags, What is a circular economy?, https://bk-bags.com/what-is-circular-economy/, (accessed July 22, 2023).
10. S. Geisendorf and F. Pietrulla, *Thunderbird International Business Review*, 2018, **60**, 771–782.
11. M. Saidani, H. Kim, F. Cluzel, Y. Leroy and B. Yannou, *Proceedings of the Design Society: DESIGN Conference*, 2020, **1**, 2129–2138.
12. Y. Kalmykova, M. Sadagopan and L. Rosado, *Resources, Conservation and Recycling*, 2018, **135**, 190–201.
13. C. Vinante, P. Sacco, G. Orzes and Y. Borgianni, *Journal of Cleaner Production*, 2021, **288**, 125090.
14. H. A. Colorado, E. I. G. Velásquez and S. N. Monteiro, *Journal of Materials Research and Technology*, 2020, **9**, 8221–8234.
15. M. Saidani, B. Yannou, Y. Leroy, F. Cluzel and A. Kendall, *Journal of Cleaner Production*, 2019, **207**, 542–559.
16. A. Winterstetter, S. Heuss-Assbichler, J. Stegemann, U. Kral, P. Wäger, M. Osmani and H. Rechberger, *Journal of Cleaner Production*, 2021, **297**, 126753.
17. E. Agliardi and M. Kasioumi, *International Journal of Production Economics*, 2023, **262**, 108927.
18. J. C. J. M. van den Bergh, *Resources, Conservation and Recycling*, 2020, **160**, 104898.
19. Y. M. B. Saavedra, D. R. Iritani, A. L. R. Pavan and A. R. Ometto, *Journal of Cleaner Production*, 2018, **170**, 1514–1522.

20. L. F. Dumée, *Circ.Econ.Sust.*, 2022, **2**, 9–23.
21. L. F. Dumée, in *Circular Economy and Sustainability*, eds. A. Stefanakis and I. Nikolaou, Elsevier, 2022, pp. 359–372.
22. Pursuit of A Circular Economy & Benefits Of A Closed-Loop, https://www.cohenusa.com/blog/recycling-and-the-pursuit-of-a-circular-economy/, (accessed July 22, 2023).
23. C. Weetman, Semi-circular strategies, https://www.rethinkglobal.info/semi-circular-strategies/, (accessed July 22, 2023).
24. D. G. Bucknall, *Philosophical Transactions of the Royal Society A: Mathematical, Physical and Engineering Sciences*, 2020, **378**, 20190268.
25. Ramasubramanian, B., Tan, J., Chellappan, V., & Ramakrishna, S. (2023). Recent Advances in Extended Producer Responsibility Initiatives for Plastic Waste Management in Germany and UK. Materials Circular Economy, 5(1), 6.
26. P. Pathak, S. Sharma and S. Ramakrishna, *Materials Today Sustainability*, 2023, **22**, 100365.
27. S. Castillo-Benancio, A. Alvarez-Risco, S. Esquerre-Botton, L. Leclercq-Machado, M. Calle-Nole, F. Morales-Ríos, M. de las Mercedes Anderson-Seminario and S. Del-Aguila-Arcentales, in *Circular Economy: Impact on Carbon and Water Footprint*, eds. A. Alvarez-Risco, S. S. Muthu and S. Del-Aguila-Arcentales, Springer, Singapore, 2022, pp. 115–138.
28. R. Gómez-Prado, A. Alvarez-Risco, J. Sánchez-Palomino, M. de las Mercedes Anderson-Seminario and S. Del-Aguila-Arcentales, in *Circular Economy: Impact on Carbon and Water Footprint*, eds. A. Alvarez-Risco, S. S. Muthu and S. Del-Aguila-Arcentales, Springer, Singapore, 2022, pp. 139–159.
29. I. Directorate-General for Internal Market, S. Bobba, P. Claudiu, D. Huygens, P. Alves Dias, B. Gawlik, E. Tzimas, D. Wittmer, P. Nuss, M. Grohol, H. Saveyn, F. Buraoui, G. Orveillon, T. Hámor, S. Slavko, F. Mathieux, M. Gislev, C. Torres De Matos, G. A. Blengini, F. Ardente, D. Blagoeva and E. Garbarino, *Report on critical raw materials and the circular economy*, Publications Office of the European Union, LU, 2018.
30. R. Navarro, The Carbon Emissions of Big Tech, https://www.electronicshub.org/the-carbon-emissions-of-big-tech/, (accessed July 22, 2023).
31. S. C. Neubauer, *Ecosystems*, 2021, **24**, 2079–2089.
32. A. R. Mosier, A. D. Halvorson, G. A. Peterson, G. P. Robertson and L. Sherrod, *Nutr Cycl Agroecosyst*, 2005, **72**, 67–76.
33. K. P. Shine, J. S. Fuglestvedt, K. Hailemariam and N. Stuber, *Climatic Change*, 2005, **68**, 281–302.
34. Energies | Free Full-Text | Energy Consumption under Circular Economy Conditions in the EU Countries, https://www.mdpi.com/1996-1073/15/21/7839, (accessed July 19, 2023).
35. L. Laureti, A. Costantiello and A. Leogrande, 2022.
36. A. M. Khan and M. Osińska, *Energies*, 2022, **15**, 7839.
37. J. C. Carpio-Aguilar, J. Rincón-Moreno and M.-L. Franco-García, in *Sustainable Development Goals and Sustainable Supply Chains in the Post-global Economy*, eds. N. Yakovleva, R. Frei and S. Rama Murthy, Springer International Publishing, Cham, 2019, pp. 225–240.
38. Ramasubramanian, B., Sundarrajan, S., Rao, R. P., Reddy, M. V., Chellappan, V., & Ramakrishna, S. (2022). Novel low-carbon energy solutions for powering emerging wearables, smart textiles, and medical devices. Energy & Environmental Science, 15(12), 4928–4981.
39. F. Di Maio, P. C. Rem, K. Baldé and M. Polder, *Resources, Conservation and Recycling*, 2017, **122**, 163–171.
40. H. Khayyam, M. Naebe, A. S. Milani, S. M. Fakhrhoseini, A. Date, B. Shabani, S. Atkiss, S. Ramakrishna, B. Fox and R. N. Jazar, *Energy*, 2021, **225**, 120113.
41. A. Tukker, *Journal of Cleaner Production*, 2015, **97**, 76–91.
42. K. L. Huff, thesis, Purdue University Graduate School, 2021.
43. R. Lantto, V. Järnefelt, M. Tähtinen, A.-S. Jääskeläinen, J. Laine-Ylijoki, A. Oasmaa, H. Sundqvist-Andberg and N. Sözer, *Industrial Biotechnology*, 2019, **15**, 12–19.
44. J. Pitt and C. Heinemeyer, in *Environment, Ethics and Cultures*, Brill, 2015, pp. 245–260.
45. M. Borrello, L. Cembalo and V. D'Amico, *Agric Econ*, 2023, **11**, 19.

46. V. Parida, T. Burström, I. Visnjic and J. Wincent, *Journal of Business Research*, 2019, **101**, 715–725.
47. Circular Economy, https://www.circular.sg/circular-economy, (accessed July 22, 2023).
48. F. U. Rehman, B. M. Al-Ghazali, A. G. Haddad, E. A. Qahwash and M. S. Sohail, *Sustainability*, 2023, **15**, 5181.
49. O. Okorie, J. Russell, R. Cherrington, O. Fisher and F. Charnley, *Resources, Conservation and Recycling*, 2023, **189**, 106756.
50. T. A. Kurniawan, M. H. Dzarfan Othman, G. H. Hwang and P. Gikas, *Journal of Cleaner Production*, 2022, **357**, 131911.
51. Q. Liu, A. H. Trevisan, M. Yang and J. Mascarenhas, *Business Strategy and the Environment*, 2022, **31**, 2171–2192.

Chapter 2
Life Cycle Assessment and Tools

Abstract This chapter offers an overview of Life Cycle Assessment (LCA) as a crucial tool for evaluating the environmental impact of products, processes, and systems. The chapter begins by discussing the benefits of using LCA and its potential to inform decision-making processes and identify opportunities for improvement. The chapter then covers the different types of LCA available and provides guidance on how to choose the appropriate type of LCA for a particular study. The steps involved in LCA are presented in detail, including goal and scope definition, inventory analysis, life cycle impact assessment, and interpretation of results. Specific aspects of LCA are also discussed, such as linear model life cycle inventory and inventory modelling. The chapter emphasizes the importance of data availability and integrity, including the temporal, geographic, and technological coverage of datasets and the use of open-source and subscription databases. Additionally, materials inflow and outflow analysis are presented as an essential aspect of LCA. The chapter concludes by highlighting the standards for LCA and Materials Circular Economy (MCE), providing readers with a comprehensive guide to LCA and its role in promoting sustainable practices.

Keywords Cradle to grave · Inventory analysis · Databases · Materials flow analysis

2.1 Introduction to LCA

Life cycle analysis (LCA) is a systematic method for assessing the environmental effect of a product, process, or activity across its full life cycle, from the extraction of raw materials to disposal at the end of its useful life [1]. LCA is a powerful tool for sustainability decision-making since it identifies areas where changes may be done to decrease environmental effect. The examination of a product's ecological impact in LCA) is centred on its whole product lifecycle, which includes numerous stages such as extraction of raw materials (Cradle), production, processing, shipping, consumption, sale, and waste disposal (Grave) [2]. The particular phases to include

S. Ramakrishna and B. Ramasubramanian, *Handbook of Materials Circular Economy*,
https://doi.org/10.1007/978-981-97-0589-4_2

in the evaluation are determined by the amount of data available and the goal of the study. There are 4 phases in a LCA analysis of a material or product system [3].

Phase 1: The extraction of raw materials is the initial stage of a product's lifecycle. Mining, drilling, lumbering, or any other sort of resource extraction necessary to manufacture the product must be included. The environmental effect of this step is determined by the type of resource being harvested, the extraction method utilized, and the resource's geography [4].

Activity 2.0

Think about all possible resource being harvested, the extraction method utilized, and the resource's geography for a making a pet plastic 1l water bottle.

The resources that can be harvested to make plastic products include Crude oil, natural gas, monomer productions, Biomass etc.

Possible extraction methods include drilling for crude oil and natural gas, fracking for natural gas, mining for coal and harvesting of plant-based resources.

Geography of the resource extraction will also depend on the type of resource being harvested, but for this case, favorable geography include:

The Middle East, North America, and Russia for crude oil extraction.
The United States, Canada, and Russia for natural gas extraction.
China, India, and the United States for coal mining.

Phase 2: Following the extraction of raw materials, they are treated and produced into the finished product. This stage includes a variety of procedures such as refining, production, assembling, and packaging. This stage's environmental effect is determined by the manufacturing technique employed, the energy source used to power the production, and the waste created throughout the manufacturing process [5].

Phase 3: After being manufactured, the product is delivered to end consumers and utilized by them. This stage comprises the shipment of the good to retail outlets, the usage of the product by the customer, and any maintenance necessary while the product is in use. The environmental effect of this step is determined by the mode of transportation utilized, the energy consumed during product usage, and the maintenance needs.

Phase 4: When a product's useful life is ended, it is discarded or recycled. This step comprises product collection, transportation, and disposal or recycling. The environmental impact of this step is determined by the method of disposing employed, the amount of energy consumed during the disposal process, and the efficiency of the recycling process [6]. The LCA incorporates the processes as shown in Fig. 2.1.

The choice of the specific phases to include in the assessment depends on the level of data availability and the purpose of the analysis.

Cradle-to-grave, cradle-to-gate, and cradle-to-cradle are the three basic life cycle models that is utilized in LCA, as shown in Fig. 2.2. Cradle-to-gate only evaluates a product's environmental effect until it leaves the production and is transferred to the customer, which encompasses phases 1 and 2 of the life cycle. Cradle-to-grave,

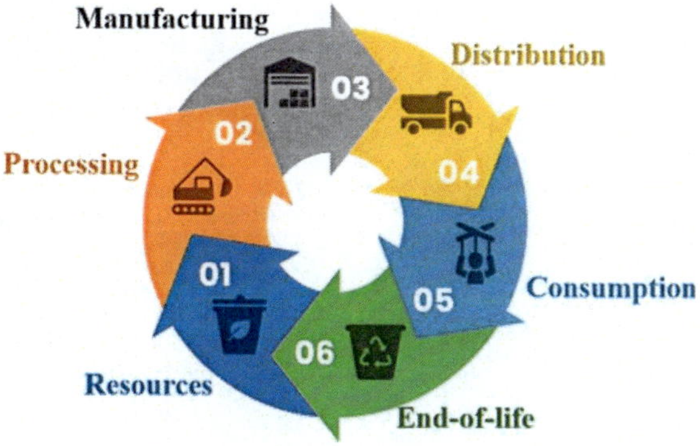

Fig. 2.1 Stages involved in cradle to cradle

on the other hand, encompasses all four phases of a product's life cycle, offering a comprehensive assessment of the product's impact from conception to disposal. Cradle-to-cradle is a variant on cradle-to-grave in that it incorporates a recycling/upcycling phase in place of the trash disposal step. The objective is to reduce the demand on extracting raw materials and trash disposal by developing technologies. This strategy seeks to establish an MCE in which goods are developed with the ultimate objective in sight and waste is reduced [8].

2.2 Benefits of LCA

LCA is a way of assessing a product's or service's environmental effect over the course of its full life cycle. One of the primary advantages of LCA is that it assists in identifying possibilities to lessen a product's or service's environmental effect. LCA identify areas where energy or resource savings can be made by assessing the whole life cycle of a product or service, such as decreasing energy or water use, eliminating waste, or procuring materials from more sustainable sources. LCA promotes transparency and accountability by giving stakeholders with information on the environmental implications of a product or service. This can aid in the development of trust and credibility among consumers, investors, and other stakeholders [9]. LCA also helps decision-makers reach better conclusions by giving insight into the ecological implications of various alternatives. Lastly, LCA spurs innovation by discovering new potential for eco-friendly product and process design. Designers and engineers could build new, more sustainable procedures that decrease environmental effect while still fulfilling client expectations by comprehending the outcomes of LCA [10].

Fig. 2.2 Three basic life cycle models for LCA [7]

2.3 Types and Choice of LCA

The techniques are characterized as follows in the Shonan database standards glossary, Sonnemann, G., & Vigon, B. (2011). *Global guidance principles for Life Cycle Assessment (LCA) databases: a basis for greener processes and products.* United Nations Environment Programme. The attributional life cycle assessment (ALCE) technique assesses the complete sentient production system and its total environmental effect [11]. ALCA then attributes a portion of this influence to specific goods and their life cycles. The evaluated functional unit may encompass many product functions, and the total environmental effect may be estimated by aggregating the environmental impact of all product life cycles. ALCA's additivity rule has a substantial impact on the product system. This distinction does not imply that ALCA ignores the effects of a changing environment, but rather that the functional unit or product

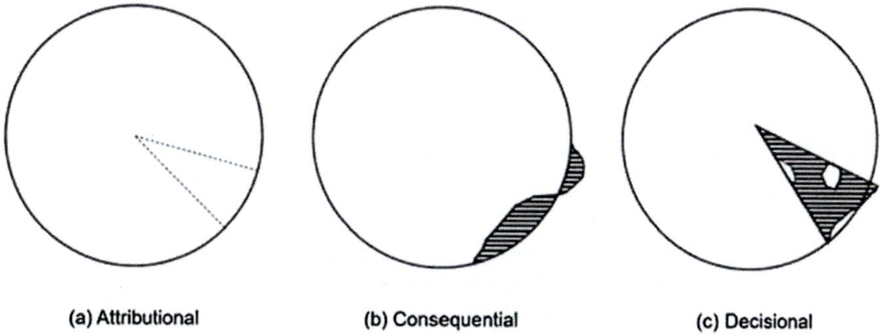

Fig. 2.3 Types of LCA [15]

life cycle does not cause the change. The ALCA model assumes that the system boundaries, functional unit, and allocation method remain constant over time and that the environmental impacts are directly proportional to the number of inputs and outputs [12]. The CLCA method, on the other hand, examines how a product selection or demand shift affects the worldwide environmental effect [13]. This method evaluates how the choice affects the overall human/industrial system and its worldwide environmental effect, as seen by the deeper marking in the right circle of Fig. 2.3. For example, if the introduction of a new product leads to changes in consumer behavior, such as increased demand for energy or water, these impacts are also considered in the assessment. The most commonly used attributional LCA model is the ISO 14040/44 standard [14].

The distinction between ALCA and CLCA. The rings indicate worldwide environmental trades as a whole. attributional LCA aims to clip off the section with dotted that belongs to a certain human influence in the left circle. In the right circle, CLCA seeks to record the variation in ecological interactions caused by the addition or removal of a single human activity [14].

Apart from ALCA and CLCA, the third type is Social LCA. This method evaluates the social and socio-economic impacts of a product or process throughout its life cycle, such as its impact on human health, well-being, and labor rights. It is often used to identify the social hotspots of a product or process, such as human rights violations or poor working conditions (Fig. 2.4). SLCA is suitable for organizations that want to understand and improve their social and socio-economic performance [16].

2.4 Steps Involved in LCA as Per ISO14040 and ISO14044

There are four basic steps involved in LCA analysis as shown in Fig. 2.5.

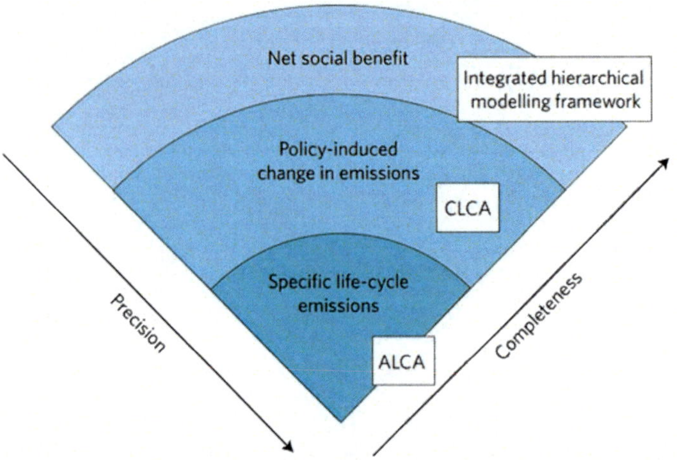

Fig. 2.4 Example model for ALCA, LCA may be used to calculate particular life-cycle emissions for certain system limits. CLCA is appropriate for analyzing legislation changes in emission, but it must cope with large uncertainty [16]

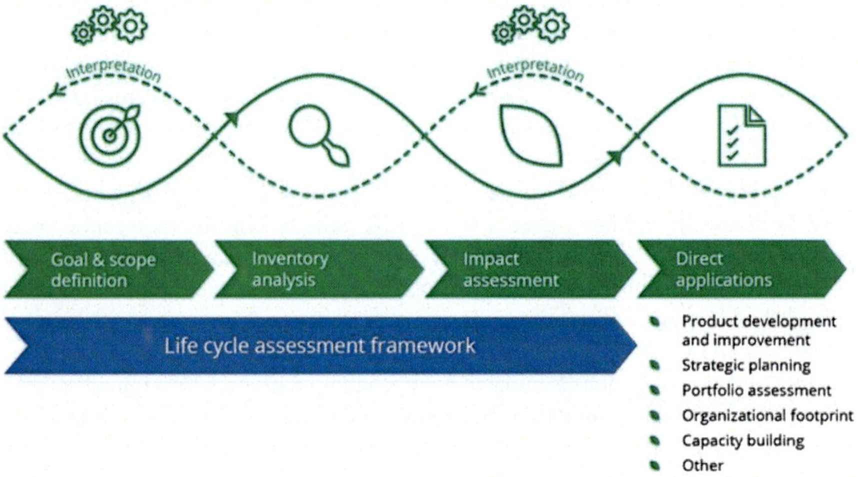

Fig. 2.5 Steps in LCA as per ISO standards [17]

(i) Goal and scope definition: This step involves defining the goal and scope of the LCA study. The goal defines the purpose of the study, while the scope defines the system boundaries and the functional unit. The functional unit is a quantitative measure of the performance of the system being studied. For example, if the system being studied is a car, the functional unit could be defined as the distance travelled or the number of passengers transported. The scope also includes the life cycle stages that will be considered in the study, such as raw material extraction, manufacturing, use, and disposal. Finally, the impact categories and the data quality requirements are also defined in this step [18]. The LCA process involves four main steps [19].

(ii) Inventory analysis: In this step, the data on the inputs and outputs of the system being studied are collected and quantified. This includes the raw materials used, energy consumption, emissions to air, water, and soil, and waste generated. The data collection can be done through surveys, measurements, or secondary sources such as databases and literature. The collected data are then organized and compiled in a life cycle inventory (LCI), which is a comprehensive list of inputs and outputs of the system at each life cycle stage.

(iii) Impact assessment: The impact assessment step involves assessing the environmental impacts of the system being studied. The LCI data are analyzed using established impact categories, such as global warming potential, acidification potential, eutrophication potential, and human toxicity. Each impact category is associated with a set of characterization factors, which convert the LCI data into impact scores. The impact scores are then aggregated into a single score for each impact category and compared to relevant environmental benchmarks or threshold values.

(iv) Interpretation: The final step in LCA involves interpreting the results of the impact assessment and drawing conclusions. The interpretation step includes identifying the significant contributors to each impact category, evaluating the data quality and uncertainties, and performing sensitivity and scenario analyses. Based on the results, recommendations can be made for improving the system's environmental performance, such as changing the design, materials, or energy sources used, or modifying the life cycle stages considered. Finally, the results and conclusions are communicated to stakeholders through a report or other communication means [20].

2.4.1 Goal and Scope Definition

The goal and scope of an LCA analysis are crucial for ensuring the accuracy and usefulness of the results. To achieve this, there are several considerations and precautions that should be kept in mind when defining the goal and scope. Firstly, it is important to clearly define the goal of the LCA analysis and ensure it is relevant to the decision-making context. This will ensure that the study is focused, and the results are useful. Additionally, the functional unit should be defined, which is a quantifiable measure of the product or service being analyzed. It should be representative of the product or service and allow for comparison with other products or services. The system boundaries should also be specified to define the extent of the study and what processes and inputs will be included. This includes the entire life cycle of the product or service, from cradle-to-grave, or a partial life cycle, such as from raw material extraction to production. This will help ensure that all relevant processes and impacts are considered in the analysis. Furthermore, it is important to choose appropriate impact categories based on the decision-making context and the goals of the study. This includes considering both environmental and social impacts. Additionally, the data used in the analysis should be accurate, reliable, and representative of the processes being analyzed. This includes data on energy use, raw materials, emissions, and waste. Uncertainties and limitations should also be considered in the LCA analysis. This involves acknowledging and addressing these limitations, such as through sensitivity analyses and the use of conservative assumptions where data is uncertain. Double-counting impacts that occur in multiple stages of the life cycle should also be avoided, as this can lead to inaccurate results and misleading conclusions. Lastly, it is important to be transparent and clearly document and communicate the goal and scope of the LCA analysis, as well as the methods and data used. This will ensure transparency and credibility of the results and allow for others to review and replicate the study [21].

Here is an example of "Defining Goal and Scope" of an Aluminium Foil manufacturing company. We conducted an initial small scale industrial survey on few SMEs in India, Singapore, and China to validate the recommended GS model and here are the findings, reported after obtaining full copyrights with permission from the SMEs.

Clearly define the goal: The goal of the LCA analysis is to assess the environmental impacts associated with the production of aluminum foil in order to identify opportunities for improvement and guide decision-making.

Define the functional unit: The functional unit for this study is one tonne of aluminum foil produced.

Specify the system boundaries: The system boundaries for this study include the entire life cycle of the aluminum foil, from raw material extraction to end-of-life disposal. This includes the following stages: bauxite mining, alumina refining, primary aluminum smelting, aluminum rolling and finishing, transportation of raw materials and finished products, and end-of-life disposal.

Choose appropriate impact categories: The impact categories chosen for this study are greenhouse gas emissions, energy consumption, water use, land use, and solid waste generation. These impact categories were chosen based on the decision-making context and the goals of the study.

Use appropriate data: The data used in the LCA analysis will be based on industry averages and specific data from the company's operations. This includes data on energy use, raw materials, emissions, and waste.

Consider uncertainties and limitations: The LCA analysis considered the uncertainties and limitations associated with the data used and assumptions made in the study. For example, to assess the impact of different bauxite mining methods, we vary the data inputs used in the LCA study. For example, we can assume that the primary aluminium is sourced from:

Conventional bauxite mining with a high-impact score.
Conventional bauxite mining with a low-impact score.

Sustainable bauxite mining with a low-impact score.

Avoid double-counting: Double-counting of impacts will be avoided by using appropriate allocation methods when multiple products are produced from the same process or material.

Transparency: The goal and scope of the LCA analysis, as well as the methods and data used, will be clearly documented and communicated to ensure transparency and credibility. The study will be conducted in accordance with ISO 14040/44 standards for LCA [22].

2.4.2 Inventory Analysis

The economy-environment system refers to the interactions between the economy and the environment. It involves the production and consumption of goods and services, which can have both positive and negative impacts on the environment. In the context of a product's LCA, upstream refers to the environmental impacts associated with the production and transportation of the raw materials and components used to create the product. Downstream refers to the environmental impacts associated with the product's use, maintenance, and disposal. Examples of upstream impacts might include the environmental damage caused by mining and processing raw materials, such as metals or fossil fuels, used in the production of a product. Downstream impacts might include the environmental effects of the product's use, such as emissions from a car's exhaust, or the waste generated from a product's disposal, such as landfill or incineration [23]. In an inventory analysis, flow diagrams are used to create a detailed inventory of all the inputs and outputs associated with a product or process. Firstly, the flow diagram should be kept simple and easy to read, with appropriate scales used for the inputs and outputs. Additionally, it should clearly show the direction of flow between different stages of the product or process, and use standardized symbols to represent inputs and outputs. To ensure transparency and replicability, the flow diagram should also include the source of data for each input and output. In this way, others can review and validate the analysis, ensuring its accuracy [24].

To simplify the diagram and make it easier to interpret, it is recommended to only include process boxes and economic flows represented by arrows, while excluding environmental flows and numbers.

Defining the functional unit is a crucial step when conducting a LCI analysis within a LCA study. The functional unit serves as a measurable unit that defines the reference point for comparing alternative products or systems, and it can take the form of either a physical unit, such as weight or volume, or a functional unit, such as the service provided, or the amount of energy generated. The functional unit is important for a multitude of reasons. Firstly, it ensures that all products or systems being compared are evaluated on the same basis, providing a clear and consistent foundation for comparison. Additionally, the functional unit selection impacts the quantity of data required for collection. For example, if the functional unit is one ton of product, then data collection will involve inputs and outputs related to the production of one ton of the product [25].

Moreover, the functional unit assists in identifying trade-offs between environmental impacts, as comparing alternative products or systems based on the same functional unit allows for the identification of relative benefits and drawbacks. For instance, a product may have lower greenhouse gas emissions per functional unit but require more water for production. By assessing the relative importance of these impacts, analysts can weigh the trade-offs between environmental consequences.

For LCA analysis, data can be obtained in two ways: either by creating a database or by importing it from the cloud, hardware, or the internet. In this instance, we will focus on the latter and explore how to import data using online tools. The Nexus Open LCA Database is an excellent example of a readily accessible and easy-to-use tool. Nexus LCA Database [26]. Nexus LCA incorporates data from world-class LCA data sources such as the ecoinvent center (ecoinvent database), PE International (GaBi databases), and the European Commission's Joint Research Centre (ELCD database). Nexus datasets may be simply loaded into the openLCA program. The openLCA and Nexus databases share a set of fundamental flows and other reference data that have been harmonized in collaboration with the respective data sources to resolve methodological discrepancies, such as waste modeling. Some of the database in Nexus LCA are open source like exiobase, OzLCI2019. Eco Invent is a subscribed database [27].

Inventory analysis is a crucial process for understanding the resource consumption and environmental impact of a product or process. It involves analyzing the inputs, outputs, and total requirements of the system being studied. Inputs refer to all the materials, energy, and other resources that are required for the production and operation of the system being analyzed. Inputs are typically divided into three categories, namely direct inputs, indirect inputs, and embedded inputs. Direct inputs are the materials and resources that are directly used in the production or operation of a product or process. Indirect inputs are the materials and resources that are required to produce the direct inputs, and embedded inputs are the resources used in the production and operation of the system but are not physically part of the final product or process. Direct outputs are the tangible products or materials that are produced by the system being analyzed, such as the final product, by-products, or waste. Indirect outputs, on the other hand, are emissions or waste generated by the production and use of the direct outputs. These can include things like greenhouse gas emissions, air and water pollution, and waste disposal. Understanding and quantifying all three types of inputs is important in conducting a comprehensive inventory analysis to determine the resource consumption and environmental impact of a product or process. Outputs refer to all the products, emissions, and waste generated by the system being analyzed. Outputs are typically divided into two categories, namely direct outputs and indirect outputs. Total requirements are the sum of all the inputs required to produce a unit of the product or service being analyzed [28].

Inputs and outputs are typically defined and quantified based on their name, category, subcategory, and amount. The name of the input or output refers to the specific material, energy, or service being analyzed. Categories and subcategories are used to group inputs and outputs into broader environmental impact categories. The amount of inputs and outputs refers to the quantity of each material or resource used or produced by the system being analyzed, typically expressed in physical units such as kilograms or litres. By defining inputs and outputs based on their name, category, subcategory, and amount, it is possible to conduct a detailed and comprehensive analysis of the environmental impacts of the system being studied. This approach enables the identification of key areas of resource consumption and environmental impact, allowing for targeted efforts to improve the sustainability of the product or process [29].

2.4.3 Linear Model Life Cycle Inventory

LCI models are typically simple and straightforward, with each process fixed in time and considered as part of a larger, homogenous system. However, when considering the consequences of a particular product, things become more complicated. The specific technologies available and the competitive market environment can have a significant impact on the final outcome. Therefore, consequential LCI models often incorporate more complex models, such as those that account for economic equilibrium, agent behavior, or dynamic changes over time [30]. While the goal of consequential models is to show how different activities affect the environment and each other, they often need to include scenarios to represent various possible outcomes. Despite the theoretical appeal of this approach, it can be challenging to make the consequences of decisions clear and unambiguous. For example, even if the consequences of a specific case are defined, a new technology could replace multiple existing ones, making the impact of decisions difficult to predict [31].

2.4.4 Establishing Limits in an Inventory Model with an Unlimited Supply

Establishing boundaries is a critical step in LCA and LCI modelling as it enables a focus on significant parts while avoiding insignificant supply chains. Defining system boundaries explicitly is also essential for impartial product comparisons. Challenges with boundaries exist both within and between technosphere and biosphere. Two common approaches to address technosphere boundaries are quantitative cut-off rules and related data collection. Quantitative cut-off rules set a threshold that excludes products and their upstream supply chain which do not meet a certain quantitative contribution. The threshold delimits the overall investigated system, with amounts reducing while going backward in the supply chain. However, if products have different units, the quantitative cut-off may not reflect the complete quantitative contribution of flows to the inventory result. In a well-specified system, as one moves backward in the supply chain, flows grow smaller, and loops converge [32]. Aluminium manufacturing, for example, requires some aluminum but not more than it generates. Similarly, maize production necessitates the consumption of a certain amount of corn as seed, but less than it yields. Items that do not deliver a particular quantity over the threshold are removed from the system and their supply chain. A threshold determines the overall system size as quantities get smaller for operations further away from the functional unit. The threshold value is usually determined by the amount of energy or materials in the production. When items have various units, such as kilos of wheat against liters of petroleum, or the number of steel plates in kilograms versus a vehicle component, the quantifiable cut-off may be insufficient to reflect the precise statistical contribution of the inflows to the overall inventory outcomes [33].

The ISO 14040 standard recommends that cut-off thresholds be measured in terms of relative contribution to environmental impacts since each flow amount can only estimate the impact it has. However, when building a product system and deciding whether to add a new process, the entire environmental impact of the supply chain is usually unknown, and therefore cannot be utilized as a threshold. Another form of cut-off rule eliminates processes or flows that fall under a specific type or classification, such as infrastructure, as research shows that they usually do not make a substantial contribution to the results of an LCA. Furthermore, some LCA studies concentrate only on certain steps in the life cycle while excluding others, such as cradle-to-gate analyses that only account for the life cycle until the product leaves the producer. However, this is not generally classified as a cut-off and will not be explored further. The application of cut-off criteria must be specified during the goal and scope definition phase of the LCA. Although they can be beneficial in creating models and concentrating efforts on the critical parts of the life cycle, many LCA software systems do not support them directly, and they are often subject to approximation due to differing units in databases. The cut-off criterion can be applied ex-post to ensure the quality of previously developed systems. This is required by some Environmental Product Declarations and Environmental Footprint Category Rules, which stipulate that the quantity of excluded materials or environmental impact must not exceed a certain threshold. However, this threshold can only be determined once the complete system is known [34].

ISO 14040 defines an ideal scenario where no threshold or cut-off is necessary. The system boundary should define the unit processes included in the system. Defining the boundary between the technosphere and biosphere is particularly challenging, especially for renewable resources. Deciding whether agricultural soils belong to the technosphere, or biosphere is often tricky. While they are essential for technical activities, they provide supporting ecosystem services. The definition of system boundaries in agricultural production systems is critical and significantly impacts results. Complex soil dynamics, such as erosion and nutrient leaching, also affect the meaning. Achieving carbon net sequestration or emissions depends on changes in soil management type until a new equilibrium level of carbon in the soil is reached.

Additionally, the status of carbon in the soil and carbon flows is influenced by the history and location of the soil under study [35].

2.4.5 Creating Models for Specific Geographic Areas

In order to construct an accurate model, it is crucial that the location where a product is produced and the location where it is consumed are identical, except when they are connected through a transportation service. For example, if electricity from Norway is intended to be used in Italy, it must be transported there first. Moreover, processes and the corresponding flows can differ depending on the location, particularly if they are affected by natural factors. For instance, a photovoltaic cell's energy yield varies based on aspects such as altitude and latitude. Additionally, the same consumption of resources or discharge of pollutants can have varying consequences depending on the area, such as the availability of resources or the state of air and water bodies. For instance, releasing particulate matter in urban regions where more people are exposed has a larger effect, while withdrawing water in dry regions such as the Arabian Peninsula has a more significant impact than drawing the same amount in regions with ample water, such as the Netherlands. The necessary level of detail and precision in the spatial aspect depends on the scale of the effects, whether they are local, regional, or global [36].

2.4.6 Spatial Archetypes

There isn't a method for representing geographic locations in LCI that is now commonly accepted. Instead, a variety of techniques are applied, depending on the LCA databases and tools in use. One strategy that is frequently used in many LCI databases, including ecoinvent and International Lifecycle Data System/Product Environment Footprint (ILCD/PEF), is to subdivide flows into different categories. With the help of this technology, it is possible to pinpoint the locations of emissions, such as high- or low-population areas, agricultural, industrial, or forestry zones, or certain kinds of water bodies like lakes, groundwater, rivers, or fossil water. This method, known as "spatial archetypes," can improve the categorization of flows connected to human health [37].

Flows are typically divided into sub-compartments based on their spatial properties using spatial archetypes. This technique is used to pinpoint the actual location where emissions take place and to more exactly and thoroughly describe the environmental effects of such emissions. The spatial archetypal approach divides flows into several spatial compartments based on diverse elements like population density, kind of land use, and type of water body. For instance, the kind of water body, such as a lake or river, and the location of the emission, such as a heavily populated urban area or a rural agricultural region, can be used to categorize a flow of pollutants into a water body. Moreover, an LCA database can be used to regionalize or nation flows. This entails supplying unique elementary flows for specific nations or areas, which are commonly identified using ISO two- to three-letter identifiers and may be enhanced or updated by different databases (such as CH for China, RER for rest of Europe, and RNA for rest of North America in ecoinvent). This method can be used and applied to a variety of procedures. For example, the AWARE water footprint method employs a flow-based regionalization strategy and country-specific water scarcity characterisation criteria. Because to the huge variations in water availability across the nation, this method may not be able to precisely predict the effects of water removal at a specific site for big countries like China. China's average water availability is 40 m^3/m^3, but certain areas, like Hunan, see substantially greater precipitation rates. A more accurate characterization factor of approximately 0.4 m^3/m^3 is obtained for the Hunan province by incorporating geographic information watershed level categorization variables provided by the AWARE technique. This results in a much low specific prediction of 0.3 m^3 rather than 25 m^3 while using 0.5 m^3 of water [38].

Regional locations are often only modelled for the forefront system during an LCA, while the backdrop system uses generic locations taken from databases. Notwithstanding, it is possible to use a site-specific technique for modelling general locales, which is important for large models. In addition to specifying the supported inventory and assessment methods, the purpose and scope of a LCA also provide the terminology for flows to guarantee uniformity, such as differentiating between various forms of dust. A dataset's integrity is also assessed according to whether it contains all pertinent flows with the proper nomenclature, providing an even balance between energy and mass. While it is simple to spot inaccurate terminology, it can be more difficult to verify that all concerned and emissions are accurately estimated [39].

2.4.7 Advanced Inventory Modelling

Advanced Inventory Modelling (AIM) is a methodology used in LCA to develop detailed inventories of materials, energy, and emissions involved in the life cycle of a product or service. AIM allows for a more comprehensive and accurate assessment of environmental impacts by accounting for factors such as regional differences, dynamic processes, and uncertainties [40].

Navigating the Complexities of Multifunctional Modelling

Multifunctionality is a major difficulty in LCI modelling that results from the co-production of various functions in a single process. The ISO has suggested a step-by-step process to deal with this problem, which entails either breaking the unit process down into smaller processes or enlarging the bounds of the product system to incorporate more functions associated with the co-products. This strategy makes sure that all relevant information about the environmental load is acquired and that the co-production is appropriately reflected in the model. The ISO advises using allocation if the first stage is not practical. This entails dividing up the environmental costs associated with each product in accordance with their fundamental physical connections, such as weight or energy content. LCI modelling provides a significant problem in selecting an acceptable method for each scenario, and it is crucial to make sure that the method portrays the processes' functionalities appropriately. It is emphasized the question's classic nature by pointing out that it has long been a problem in LCI modelling. In general, good multifunctionality modelling is essential for obtaining trustworthy results in LCI investigations [41].

Three methods for creating an LCI model are described in the ISO standard, each with a unique practical application. The first strategy is system subdivision, which calls for more work to improve data collecting and narrow the focus to the study's subject matter. Subdivision might not always be possible, particularly if the processes are not autonomous in terms of both geographical location and economics. The second approach, system expansion, involves system enlargement or the avoided burden strategy, although it can result in a larger, more complex model that needs more data. The third strategy is allocation, which is frequently debatable and has several principles, but there isn't a single way that offers a usually palatable resolution. In summary, the choice of which method to use to deal with multifunctionality should be decided in the aim and scope description because it may have a big impact on the final LCI model [42].

Real-world manufacturing processes and product life cycles are time-dependent, with each process characterized by a certain temporal dimension. Although the LCI model's core is time-agnostic, it seeks to capture a dynamic, time-dependent technosphere. Results from LCI are expressed as absolute values rather than as physical flows with timing information. The use phase of products that require maintenance operations, technical maturity, and seasonal processes are just a few of the ways that time affects LCI modelling. However, some time-related factors, such as the storage duration and constantly increasing or falling stocks, can be ignored by LCI modelling. It is necessary to imagine an emerging technology in a more developed future state when comparing it to more established technologies.

Time plays a crucial role in various stages of a product's life cycle. For instance, certain products like buildings and cars require maintenance procedures during the usage phase. Over the course of the product's existence, maintenance actions are normally anticipated at pre-set intervals. Alongside input and output, a preservation phase could also happen, which could change the stock. In economics, a storage term is used to explain this impact; however, in LCI modelling, this word is omitted. Second, LCA frequently fails to take into account the shifting input and output flows that seasonal processes—such as agricultural systems—have throughout the year. Finally, the technological maturity of the various technologies examined in LCA may have varied. Early on in the history of LCA, the research subjects were goods that had been manufactured and used in society for a very long time [43].

Additionally, during a life cycle, resource consumption and outputs may occur at various phases and at various times, occasionally with long pauses of several years or even centuries. These gaps in time give rise to questions about justice in terms of both present and future effects, as well as justice between generations. From a methodological perspective, the impact of long-term emissions may be overestimated if the same characterisation criteria are applied to short-term and long-term emissions. A huge amount of pollution released all at once may not have the same effect as the same amount released gradually over a period of years.

High-Impact Flows and Uncertain Mechanisms

The LCI paradigm's fundamental tenet is that it is inevitable. This implies that flows at a particular time can be confidently predicted. While some experts and libraries do contain information regarding flow uncertainty, this is more often done to improve the accuracy and quality of the data in the database than to determine how likely it is that the flow will occur. LCI models only include certain, predictable flows while excluding low probability flows. According to the traditional definition of risk in threat assessments, which is equal to the likelihood of occurrence times the possible impact, these low likelihood flows may be regarded as a risk if they have an effect. The fundamental LCI model is reliable. Flows are modelled with certainty at an unspecified time and location. Some practitioners and databases include uncertainty information in FOW (Future of Work); however this is done primarily to address the dependability and data quality of the information rather than the chance of the FOW (Future of Work) happening at all. Only deterministic, completely definite fows are recorded in an LCI model, whereas low probability fows are eliminated. If these fows have an influence, the impact may be defined as risk using the conventional definition in risk assessment, where risk equals probability time impact. While LCI's sensitivity analysis can handle particular incidents by changing flows, such as raising or lowering specific emissions with a probability, this method is constrained in its capacity to model other choices beyond complete inventories. Moreover, it is unsuited to probabilistic modelling of chains of effects. To address concerns about the dependability of technical plants, particularly nuclear power plants, other approaches, including Bayesian networks used for risk assessment and failure mode effect studies, are more suitable. Bayesian networks shouldn't be viewed as the only tool for these tasks, though [44].

Recent research on the integration of risk assessment and life cycle assessment (LCA) can be classified into three main clusters. The first category pertains to site-specific assessments, which begin with an ecological input–output analysis, aimed at evaluating distinct potential hazards in different locations. The second category focuses on expanding the scope of risk analysis beyond specific production phases by utilizing the life cycle concept. Lastly, the third category examines the relative importance of local and global effects, particularly when concentrating on specific contexts and the issue of burden shifting. Contrarily, functional flows refer to the economic processes that determine all or part of an overall unit process's aim. This involves the waste inputs of a treatment process for waste and the product outflows, which could include services, of a production process. On the other hand, non-functional flows are those flows that are not a part of the functional flow of a unit process. Together with basic inflows and outflows, these include product inputs and waste outflows [45].

2.5 Life Cycle Impact Assessment

In the impact assessment (IA) phase, inventories data that shows emission levels and energy use must be transformed into effect categories using IA methodologies. These groups, which include several effect categories and characterization models, are also referred to as indicators and connect different LCA outcomes. Typical categories include resource scarcity, algae blooms, acidification, ecological toxicity, and global warming. Midpoint or endpoint indicators can be used to describe certain impact types. Under the climate change impact group, for example, a midway impact may be kg CO_2-equivalents/kg gas, while an endpoint impact might be the effect on ecology, such as an increase in ocean level or the global mean temperature.

An outline of resource- and emission-focused approaches to environmental indicators is given below. Resource-oriented approaches take into account all primary energy needed for a product's production, use, and disposal as well as all of the biologically productive land and sea area required to produce consumed goods and absorb generated waste. These strategies include total combined energy use and ecological footprint. The following emission-oriented methodologies for assessing environmental effect categories: CML, Eco-indicator 99, EDIP 2003, IMPACT 2002+, ReciPe, Ecological scarcity technique, IMPACT World+, ILCD 2011 Midpoint, TRACI 2.1, and LC-Impact. Moreover, the USEtox model was created expressly for comparing toxicities of products and services.

These are various environmental indicators categorized into resource-oriented and emission-oriented approaches. The former includes Cumulated Energy Use, which refers to the primary energy consumed in the production, use, and disposal of a product, and Ecological Footprint, which considers the land and sea areas required for producing consumed products and absorbing waste. Meanwhile, the latter includes CML, which uses nine baseline impact categories and twelve scientific categories to measure only midpoint impacts; Eco-indicator 99, which measures only endpoint impacts of emissions and resource categories; EDIP 2003, which measures only midpoint impacts of emission categories; IMPACT 2002+, which links 14 midpoint categories to four damage categories and is based on Eco-indicator 99 and CML 2002; ReciPe, which harmonizes midpoint and endpoint approaches of Eco-indicator 99 and CML 2002; Ecological Scarcity Method, which weighs environmental impacts based on eco-factors derived from political targets or environmental laws; IMPACT World+, which is an updated version of various LCIA methods, including IMPACT 2002+, EDIP, and LUCAS; ILCD 2011 Midpoint, which analyzes different LCIA methodologies to recommend the best method for each environmental theme; TRACI

2.1, which evaluates the impacts of goods and services in the US; and USEtox, which is a scientific consensus model for comparing the toxicity of goods and services [46].

PEF and OEF are schemes that aim to evaluate the environmental performance of products and organizations throughout their life cycle. PEF stands for "Product Environmental Footprint," while OEF stands for "Organization Environmental Footprint." These schemes were developed by the European Commission. A total of 29 pilots were conducted from November 2013 to December 2019 to evaluate the PEF/OEF. 16 of the 26 pilots—or 62% of the total—submitted their PEFCR/OEFSR to the European Commission before December 21st. By the end of January 2017, four more pilots had submitted their PEFCR/OEFSR documents, for a total submission rate of 77%. Yet, according to Kerkhof et al., two pilots were terminated in 2016 and four pilots had delays (2017).

Six steps in impact assessment as stated below,

1. Choosing the impact categories to be evaluated
2. Categorizing the individual factors that contribute to each impact category
3. Creating a model to measure the potential impact of each category using conversion factors.
4. Standardizing the potential impacts in relation to a reference point
5. Organizing the impact indicators by grouping or ranking them
6. Assigning relative weights to each impact category and providing an evaluation and report.

Standard Impact Categories: Global warming potential (GWP), Eutrophication potential (EP), Photochemical oxidation potential (POP), Acidification potential (AP), Ecotoxicity potential (ETP), Human toxicity potential (HTP), Ozone depletion potential (ODP), Land use change (LUC), Depletion of non-renewable energy resources, Water depletion potential, Soil erosion potential and Particulate matter (PM) emissions [47].

Classification: During the classification step in impact assessment, each impact category is categorized based on the types of flows and units involved. For example, in the case of the GWP impact category, the individual greenhouse gases that contribute to global warming are classified and quantified. This involves further classification into specific gases such as carbon dioxide (CO_2), methane (CH_4), chlorofluorocarbons (CFCs), nitrous oxide (N_2O), and water vapor. This means that for GWP, the impact of each specific greenhouse gas on global warming is assessed, as each gas has a different potential to trap heat in the atmosphere. By categorizing and quantifying these gases, a more accurate understanding of the total impact of a product or process on global warming can be obtained.

Characterization: During the characterization step of impact assessment, the individual elementary flows within each impact category are combined into a single indicator that represents the total impact of the product or process on that impact category. This is done by using characterization factors, which represent the potential impact of each elementary flow on the impact category. For example, in the case of the Global Warming Potential (GWP) impact category, the characterization factor for each greenhouse gas is used to convert the mass of the gas emitted into an equivalent amount of carbon dioxide (CO_2) emissions. This equivalent amount of CO_2 emissions is known as the CO_2-equivalent and is used as the single indicator for the impact category of GWP. By combining all the elementary flows into a single indicator, it becomes easier to compare the overall impact of different products or processes on the same impact category. The indicator provides a quantitative measure of the impact, which can be used to prioritize actions to mitigate the impact and to compare different alternatives [48].

Normalization: Normalization is a step-in impact assessment that involves scaling the indicator scores obtained in the characterization step to a common reference point or unit of measurement. This is done to make the results comparable across different impact categories, and to ensure that the impacts of different products or processes can be compared on a common basis. To normalize an impact category, the impact indicator score obtained in the characterization step is divided by a reference value. The reference value can be a unit of measurement, such as per kilogram of product or per unit of energy consumed, or it can be a benchmark value obtained from industry standards or best practices. For example, in the case of the Global Warming Potential (GWP) impact category, the indicator scores for each greenhouse gas are calculated in terms of their CO_2-equivalent emissions. These scores are then normalized by dividing them by a reference value, such as the GWP of one Kg of carbon dioxide. This allows the impacts of different products or processes to be compared based on their CO_2-equivalent emissions per unit of output or per unit of energy consumed.

Grouping: Grouping is another step-in impact assessment that involves categorizing the impact indicators into groups based on their similarities or differences. This can be done based on the type of impact, such as emissions to air or water, or based on the location of the impact, such as local, regional, or global. Indicators can also be grouped based on their ranking or priority level, such as high, medium, or low impact.

Weighting factor: Weighting is a step-in impact assessment that involves assigning relative values or weights to different impact categories based on their perceived importance. This is typically done through a multiple criteria analysis, which considers a range of factors such as environmental, social, and economic impacts, as well as stakeholder perspectives and priorities. The weighting step is important because it enables decision-makers to compare and prioritize different impact categories based on their relative importance. By assigning weights to each impact category, decision-makers can more effectively allocate resources and prioritize mitigation efforts to achieve the greatest overall impact reduction. For example, in the weighting step, the relative importance of each impact category would be assessed based on criteria such as the severity of the impact, the likelihood of occurrence, and the values and priorities of stakeholders. If stakeholders value the reduction of global warming potential above all other impacts, then the weighting factor for this impact category would be higher than for the other categories. This would indicate that more resources and efforts should be allocated to reducing the product's global warming potential [49].

2.6 Interpretation of Results

The interpretation involves two main steps: identification of significant issues and evaluation through completeness, sensitivity check, and consistency check. The first step, identification of significant issues, involves reviewing the results of the LCA study and identifying the environmental impacts that are most significant. This step involves a comprehensive review of the study to identify the key findings and areas of concern. The goal of this step is to determine which environmental impacts are most important and require further investigation. The second step, evaluation through completeness, sensitivity check, and consistency check, involves a detailed review of the LCA study to ensure that it is comprehensive, accurate, and reliable. This step involves several checks, including:

Completeness check: This involves ensuring that all relevant aspects of the product or process have been included in the LCA study. This includes identifying all the inputs and outputs of the product or process and assessing the environmental impacts of each.

Sensitivity check: This involves testing the robustness of the LCA study by varying assumptions and parameters to see how the results change. This helps to identify which assumptions and parameters have the most significant impact on the results and can help to improve the accuracy of the study.

Consistency check: This involves comparing the results of the LCA study to other studies or industry benchmarks to ensure that they are consistent. This helps to identify any areas where the study may be flawed or where further investigation is needed [50].

2.7 Data Availability and Integrity

2.7.1 Temporal Coverage, Geographic Coverage and Technological Coverage, Precision and Completeness

It is critical to have access to high-quality, trustworthy data while doing an LCA to guarantee that the results are accurate and relevant. An LCA can employ a variety of data sources, including confidential data from manufacturers, public databases, and literature sources. Proprietary data is information that belongs to a certain corporation or organization and is not open to the public. While this sort of data might be useful in an LCA, it can also be biased and is not always visible or independently checked. Literatures may also be a good source of data for LCAs. They can include publicly available studies, academic articles, and other sources of knowledge. While literary sources can give useful insights for LCA, it is critical and difficult to thoroughly assess the data's quality and relevancy before employing it. Regardless of the data source, it is essential to discover reliable databases that deliver high-quality, accurate data. This might entail thoroughly reviewing the data and verifying that it has been independently confirmed and validated. It is also critical to confirm that the data is relevant to the product or process under consideration and that it is compatible with other informational sources.

According to the European Commission, the EF uses a data quality assessment methodology that consists of four unique elements to evaluate the accuracy of operational datasets both in consolidated and dispersed forms. The formula's elements, which are each summed over the dataset, are the data quality requirement (DQR), technical representativeness (TeR), geographic representativeness (GR), time representativeness (TiR), and precision (P). The DQR is a crucial component of the equation and is used to establish the required data quality levels. To check that the data satisfies the DQR, the TeR, GR, TiR, and P elements evaluate the data's technical accuracy, geographical reach, temporal coverage, and accuracy, respectively [51].

$$\mathrm{DQR} = \mathrm{TeR} + \mathrm{GR} + \mathrm{TiR} + \mathrm{P}/4$$

2.7.2 *Open-Source Databases*

US LCI: The United States Environmental Protection Agency's (EPA) US LCI Database is a readily available database that offers life cycle inventory data for a wide range of materials and products. Almost 4,000 distinct datasets spanning materials, energy, transportation, and other industries are included in the database. The US LCI Database provides data on the environmental impacts of materials, processes, and products, including data on resource consumption, emissions, and waste generation. The database includes information on the entire life cycle of a product, from raw material extraction to disposal, and can be used to evaluate the environmental impacts associated with each stage of the life cycle.

Using the US LCI Database:

i. Go to the US LCI Database website (https://www.lcacommons.gov/).
ii. Create an account (if necessary) and log in.
iii. Use the browse data repositories in the US LCI Database or search in the documentation tab.

iv. Review the information provided in the dataset, including the description of the product or process, the geographical location of production, and the environmental impacts associated with each stage of the life cycle.
v. Use the data from the US LCI Database to conduct your LCA and ensure that you appropriately cite the source of the data.

OpenLCA: Nexus provides a user-friendly Open LCA interface that allows to easily access and query the database. You can search for specific products or processes, browse through different categories, or filter the data based on various criteria such as geographical location, industry sector, or environmental impact category. The database also includes a wide range of impact assessment methods that can be used to assess the environmental impacts of different products and processes.

To use OpenLCA Nexus, you will need to follow these steps:

i. Install OpenLCA software (https://www.openlca.org/download/) on your computer.
ii. Launch OpenLCA and create a new project.
iii. Click on the "Import" button and select "Nexus" as the import type.
iv. Select the desired datasets and impact assessment methods from the OpenLCA Nexus database (https://nexus.openlca.org/downloads).
v. Click on the "Import" button to import the selected data into your project.
vi. Use the imported data to perform LCA analyses.

GaBi: GaBi is a large LCA database that contains information on materials, products, and processes from a range of sectors. Researchers and practitioners use it extensively to perform LCA studies and examine the environmental effect of various goods and processes.

i. Define the goal and scope of your LCA: Before using GaBi, you need to define the goal and scope of your LCA study. This will help you determine the data you need from GaBi and ensure that you are using it in a way that addresses your research question.
ii. Access the GaBi software and database: GaBi is a paid software, so you will need to purchase a license from the thinkstep website. Once you have the software installed on your computer, you can access the database by clicking on the "Database" tab in the GaBi interface.
iii. Create a new project: To start a new LCA study in GaBi, click on "New" under the "Project" tab in the interface. This will open a dialog box where you can enter the name and description of your project.
iv. Define the system boundary: The system boundary is the set of processes and activities that you will include in your LCA study. In GaBi, you can define the system boundary by creating a process flow diagram. You can add processes and materials to the diagram by clicking on the "Processes" and "Materials" tabs in the GaBi interface.
v. Add data to your LCA model: Once you have defined your system boundary, you can start adding data to your LCA model. GaBi includes a vast amount of data on materials, products, and processes, so you can search for the data you

need by using the search function or browsing through the different categories in the database.

vi. Run the LCA calculation: Once you have added all the necessary data to your LCA model, you can run the LCA calculation by clicking on the "Calculate" button in the GaBi interface. This will generate a report that includes the results of your LCA study.

vii. Interpret and communicate your results: Once you have completed your LCA study using GaBi, you need to interpret and communicate your results. This involves analyzing the data, drawing conclusions, and communicating your findings to your audience. GaBi includes tools that can help you visualize and communicate your results effectively.

2.7.3 Subscription Databases

There are several LCA databases that require a subscription or license for access. The top 10 subscribed LCA databases based on popularity and usage are:

i. GaBi (thinkstep)
ii. SimaPro (PRé Consultants)
iii. ecoinvent (ecoinvent Centre)
iv. AGRIBALYSE (ADEME)
v. Agri-footprint (Wageningen University and Research)
vi. ELCD (European Commission Joint Research Centre)
vii. GREET (Argonne National Laboratory)
viii. USLCI (National Renewable Energy Laboratory)
ix. O-LCA (One-Click LCA)
x. EarthSmart (thinkstep).

2.8 Materials Inflow and Outflow Analysis

Material flow analysis (MFA) is an extremely effective strategy utilized by developed nations to manage complex waste streams. With applications in resource management, waste management, and environmental management, MFA has become a popular decision-support tool. Its origins can be traced back to Greek philosophers over 200 years ago. Nearly 40 years ago, Abel Wolman introduced the concept of "metabolism of cities," characterizing cities as living organisms with material and energy inputs, stocks, and outputs. According to eminent authors, MFA is an organized analysis of material flows and stocks within a specified system that links a material's sources, pathways, intermediate sinks, and final sinks. The law of conservation of matter is used to compare material balances of all inputs, stocks, and outputs to regulate MFA [52].

Dividing a process into smaller subprocesses creates a subsystem in MFA modelling. It is useful to have multiple layers of subsystems to easily comprehend the main flows and stocks on the top layer, while allowing for deeper understanding by investigating subprocesses and their connections. This is important for improving a process or solving data problems related to input/output flows.

The majority of MFA modelling methods are viewed as opaque black boxes. The quantity of materials used in a process, however, is an exception to the rule. The stock of materials, also known as the total amount of materials kept in a process, is defined in order to do this. Important parameters for describing a process include the stock's mass and its rate of change over time. Materials are kept in a final sink process when their residence time is very long. For instance, people store items for years in their homes, non-recyclables are dumped in landfills for decades, and carbon dioxide is naturally stored in the atmosphere and oceans. Stocks within a process are represented by smaller boxes inside the larger process [53].

The terms "flows" and "fluxes" have been used interchangeably in MFA modelling, but they actually have different definitions. A flow refers to the mass of a substance that moves through a conductor over time, while a flux is the flow divided by the cross-sectional area of the conductor. In MFA, a cross section is typically a person, system surface area, or entity like a household or business. Some authors use "fluxes" to describe all material exchanges between processes within a system, while others only use it to describe specific flows related to a cross section. This handbook uses a marginally distinct definition, where only flows related to a cross section are called "fluxes." This allows for easier comparison between different systems and processes [54].

Scenario	Flow 1	Cross section 1	Flux 1	Flow 2	Cross section 2	Flux 2	Flow 3	Cross section 3	Flux 3
1	Water	Pipe	kg/s	Electricity	Building	W	Natural gas	Household	m3/h
2	Cars	Road	vehicles/hour	Air Pollutants	City	µg/m3	Waste	City	kg/h
3	Oil	Tanker	t	Food	Farm	kg/year	Woodchips	Factory	t/day
4	CO2	Power Plant	t/year	Aluminum	Factory	kg/h	Plastic	Household	kg/year
5	Steel	Factory	t/year	Water	City	L/day	Paper	Office Building	kg/day
6	Chemicals	Warehouse	kg/day	Timber	Forest	m3/year	Copper	Mine	t/year
7	Plastic	Ocean	t/day	Beer	Pub	L/week	Medications	Pharmacy	units/day
8	Natural Gas	Pipeline	m3/h	Textiles	Factory	m2/h	Wheat	Country	t/year
9	Methane	Landfill	m3/h	Solar Energy	Solar panels	kW	Electronics	Household	kg/year
10	Waste	City	t/day	Beef	Farm	kg/year	Gold	Mine	t/year

Assuming k is the number of flows, p is the number of processes, and n is the number of components in a system. The mass balancing rule asserts that, after accounting for any build-up or exhaustion of components during the process, the entire mass of inflows into a process matches the entire mass of outflows from that process.

$$\sum_{Kin}^{n} minput = \sum_{Kout}^{n} moutput + mstock \tag{2.1}$$

The variable G denotes the identifier for the good, whereas the variable S stands for the identifier for the substance.

In some circumstances, it is possible to evaluate material flows using proxy data, cross-comparisons with related systems, or educated guesses. Proxy values are numbers that help with estimating or estimating the required data. Depending on the financial resources available for conducting an MFA, it may be fairly expensive to measure the real mass flows of commodities and substance concentrations. Thus, smaller systems are typically chosen for measuring flows, stocks, and concentrations, such as a wastewater treatment plant, a business, a farm, or a single private dwelling. However, because the data gathering process can be intricate and time-consuming, conducting field studies in these environments necessitates thorough preparation of the measuring approach and strategy.

The process of calculating the substance fluxes (x) caused by the flows of goods (m) can be achieved directly by using the mass flows of goods and the substance concentrations in them (c), as shown in (2).

$$\mathrm{x_m = c_{ij} + I_{ij} = I_{ij}} \tag{2.2}$$

The value of i ranges from 1 to k, while the value of j ranges from 1 to n.

It is important to note that the principle of mass balance applies to all substances in every process throughout the system, just as it does to products. The unbalanced goods can arise from missing flows, which can be eliminated by balancing the goods at the level of the system.

Typical examinations that might be carried out include the following: (1) reducing computed flow uncertainty, which results from input data uncertainty. These uncertainties should be checked and, if feasible, narrowed down if they are too large. If not, while discussing outcomes, the significant uncertainties must be accepted. (2) regulating the size of the reconciliation stage to assess the model's consistency and quality of the input data. Reconciling a lot of data suggests that there might be missing flows or inconsistent data. Reconciling marginal data typically indicates that the data and model are reliable. The average of all reconciliations divided by the average of all maximally permitted reconciliations can be used to determine the extent of the reconciliation process, particularly in well-described MFA studies. Sankey diagrams, flowcharts, and partition diagrams are typical visual aids used to convey findings. All processes, stocks, and flows must be recognized and measured.

The breadth of a flow in the graphic should be proportionate to its numerical magnitude to help the reader quickly comprehend the significance of the movement. Sankey diagrams, which is used to represent material, energy, and monetary flows, are an example of this sort of visualization that is often utilized. SFA (Substance Flow Analysis) and IOA (Input–Output Analysis) can be integrated technically to gain a more comprehensive understanding of resource and material flows in a given system or economy. SFA provides a detailed analysis of the physical flows of materials and substances within a specific system, such as a country or a production process. On the other hand, IOA provides an economic analysis of inter-industry transactions and how they relate to the overall economy. By combining SFA and IOA, researchers can gain a more complete understanding of the environmental impacts of economic activity, including the impact of material flows on the economy and the impact of economic activity on the environment. The integration of these two methods can help identify areas where changes in production processes or consumption patterns could lead to improvements in both economic and environmental performance. One way to integrate SFA and IOA is to use input–output tables as a basis for calculating material and substance flows. Input–output tables provide a comprehensive overview of the economic transactions between different sectors, including both intermediate and final products. By combining these tables with data on material and substance flows, researchers can analyze the environmental impact of economic activity and identify potential areas for improvement. [55]

2.9 Standards for LCA and MCE

One of the most widely recognized certifications is the Cradle to Cradle (C2C) certification, which evaluates products based on their environmental impact across their entire lifecycle, from production to disposal. The certification evaluates five categories, including material health, material reutilization, renewable energy and carbon management, water stewardship, and social fairness. C2C is widely used in industries such as fashion, construction, and beauty, where it helps to promote the use of sustainable materials and reduce waste [56].

Another important certification is the EU Ecolabel, which recognizes products and services that meet specific environmental criteria. The criteria are based on the product's entire lifecycle, from raw materials to production, use, and disposal. The EU Ecolabel covers a wide range of products, including cleaning products, textiles, paper products, and paints. Products that carry the EU Ecolabel are recognized as being environmentally friendly and sustainable. The Global Recycle Standard (GRS) is a certification program that recognizes products made from recycled materials. The certification evaluates the entire supply chain, from the collection of the raw materials to the production process. The GRS certification is widely used in the textile industry, where it helps to promote the use of recycled materials and reduce waste [57].

The Forest Stewardship Council (FSC) certification recognizes responsible forest management. The FSC evaluates the forest management practices and the supply chain of forest products. Products that carry the FSC certification are recognized as being sustainably sourced and environmentally responsible. The FSC certification is widely used in the paper and wood industries. In addition to certifications, there are also tools and platforms available to help companies assess the circularity of their products. The Circular Materials Assessment Tool (CMAT) helps companies to evaluate the materials used in the product, the design of the product, and the end-of-life options. The CMAT helps companies to identify areas where they can improve the circularity of their products and reduce waste. The Material Circularity Indicator (MCI) is another tool that helps companies to measure the circularity of their products. The tool evaluates the percentage of recycled or renewable materials used in the product, the recyclability of the product, and the use of environmentally friendly production methods. The MCI helps companies to track their progress towards a circular economy and identify areas for improvement [58, 59].

2.10 Conclusion

In conclusion, this chapter provides a comprehensive understanding of the key aspects of Life Cycle Assessment (LCA), including its benefits, types, and steps involved as per ISO14040 and ISO14044. The chapter highlights the importance of goal and scope definition, inventory analysis, linear model life cycle inventory, and advanced inventory modeling in conducting an LCA study. The chapter also emphasizes the significance of life cycle impact assessment and interpretation of results in providing insights for decision-making. Data availability and integrity, including temporal, geographic, and technological coverage, and the use of open-source and subscription databases, are discussed in detail. Furthermore, the chapter presents materials inflow and outflow analysis as a crucial aspect of LCA and highlights the standards for LCA and Materials Circular Economy (MCE). Overall, this chapter offers valuable insights into LCA and its role in promoting sustainable practices and informs readers on how to conduct a comprehensive LCA study.

MCQ

What is the purpose of life cycle assessments (LCAs)?

(a) To evaluate the environmental impacts of a product throughout its entire life cycle.
(b) To assess the economic viability of a product during its production phase.
(c) To analyze the social benefits of a product after its disposal.
(d) To measure the aesthetic qualities of a product during its use phase.

Answer: (a) To evaluate the environmental impacts of a product throughout its entire life cycle.

Which of the following is NOT considered in a life cycle assessment?

(a) Raw material extraction
(b) Manufacturing and production

(c) Packaging and transportation
(d) Advertising and marketing

Answer: (d) Advertising and marketing.

What is the first step in conducting a life cycle assessment?

(a) Identifying the product's end-of-life options
(b) Analyzing the energy consumption during manufacturing
(c) (Collecting data on raw material extraction
(d) (Setting environmental impact reduction targets

Answer: (c) Collecting data on raw material extraction.

Which of the following is an example of a quantitative indicator used in life cycle assessments?

(a) Visual appearance of the product
(b) (Consumer preference for the product
(c) (Amount of greenhouse gas emissions
(d) (Social media buzz around the product

Answer: (c) Amount of greenhouse gas emissions.

What is the goal of materials analysis?

(a) To determine the economic value of materials
(b) (To identify the social impacts of materials
(c) (To assess the physical properties of materials
(d) (To evaluate the environmental performance of materials

Answer: (d) To evaluate the environmental performance of materials

Which of the following factors is NOT typically considered in materials analysis?

(a) Resource depletion
(b) Toxicity and human health impacts
(c) Market demand and consumer preferences
(d) Material cost and availability

Answer: (c) Market demand and consumer preferences

What is a commonly used tool for materials analysis?

(a) Environmental Product Declaration (EPD)
(b) Life Cycle Costing (LCC)
(c) Material Safety Data Sheet (MSDS)
(d) Design for Disassembly (DfD)

Answer: (a) Environmental Product Declaration (EPD).

What information does an Environmental Product Declaration (EPD) provide?

(a) Product's carbon footprint
(b) Product's financial cost
(c) Product's durability and strength
(d) Product's aesthetic appeal

Answer: (a) Product's carbon footprint.

Which of the following statements about eco-labeling is true?

(a) Eco-labeling is mandatory for all products in the market.
(b) Eco-labels only focus on the recycling potential of a product.
(c) Eco-labeling provides standardized information on the environmental performance of a product.
(d) Eco-labels are solely based on the manufacturer's claims.

Answer: (c) Eco-labeling provides standardized information on the environmental performance of a product.

What is the purpose of conducting a materials flow analysis?

(a) To assess the social impact of material use
(b) To determine the economic value of materials
(c) To identify the sources and destinations of materials
(d) To evaluate the aesthetics of materials

Answer: (c) To identify the sources and destinations of materials.

Which of the following materials has the highest environmental impact?

(a) Recycled paper
(b) Aluminum
(c) Glass
(d) Cotton

Answer: (b) Aluminum.

Which of the following is NOT a key component of a circular economy?

(a) Reducing waste generation
(b) Promoting reuse and recycling
(c) Focusing on linear production models
(d) Designing products for longevity

Answer: (c) Focusing on linear production models.

What is the primary objective of waste minimization?

(a) To reduce the environmental impact of waste disposal
(b) To increase the profitability of waste management companies
(c) To promote the use of landfill sites for waste disposal
(d) To accelerate the decomposition of waste materials

Answer: (a) To reduce the environmental impact of waste disposal

What is the purpose of conducting a carbon footprint analysis?

(a) To measure the financial cost of a product
(b) To evaluate the social impacts of a product
(c) To assess the environmental impacts of a product's greenhouse gas emissions
(d) To determine the aesthetic appeal of a product

Answer: (c) To assess the environmental impacts of a product's greenhouse gas emissions.

Which of the following is a renewable energy source?

(a) Natural gas
(b) Coal
(c) Solar power
(d) Nuclear power

Answer: (c) Solar power.

What is the main objective of a materials selection process?

(a) To maximize material cost savings
(b) To optimize material production speed
(c) To minimize the environmental impact of materials
(d) To prioritize material aesthetics

Answer: (c) To minimize the environmental impact of materials.

What is the significance of a cradle-to-cradle approach in materials analysis?

(a) It promotes linear production models
(b) It prioritizes material disposal options
(c) It emphasizes material reuse and recycling
(d) It focuses on single-use materials

Answer: (c) It emphasizes material reuse and recycling.

Which of the following is NOT a primary indicator used in life cycle assessments?

(a) Energy consumption
(b) Water usage
(c) Material durability
(d) Land use

Answer: (c) Material durability.

What is the goal of the end-of-life phase in life cycle assessments?

(a) To assess the impacts of product disposal
(b) To evaluate the energy consumption during product use
(c) To analyze the social benefits of the product
(d) To determine the economic viability of the product

Answer: (a) To assess the impacts of product disposal.

Which of the following is an example of a sustainable material?

(a) PVC (Polyvinyl Chloride)
(b) Styrofoam
(c) Bamboo
(d) Polystyrene

Answer: (c) Bamboo.

References

1. R. Heijungs, G. Huppes and J. B. Guinée, *Polymer Degradation and Stability*, 2010, **95**, 422–428.
2. R. U. Ayres, *Resources, Conservation and Recycling*, 1995, **14**, 199–223.
3. Ramasubramanian, B., Rao, R. P., Chellappan, V., & Ramakrishna, S. (2022). Towards sustainable fuel cells and batteries with an AI perspective. Sustainability, 14(23), 16001.
4. A. Katebi, H. Soleymani Tushmanlo and G. Asadollahfardi, *Journal of Building Engineering*, 2023, **76**, 107316.
5. filipeboni, 5 Benefits of Conducting a Life Cycle Assessment, https://ugreen.io/5-benefits-of-conducting-a-life-cycle-assessment/, (accessed July 22, 2023).
6. Life Cycle Assessment (LCA) | Definition, Phases, Uses, & Example, https://www.carboncollective.co/sustainable-investing/life-cycle-assessment-lca, (accessed July 20, 2023).
7. S. R. Hinish, Cradle-to-Cradle System Design, https://www.supplychainqueen.com/cradle-to-cradle-system-design-reflections-by-dr-michael-braungart/, (accessed December 15, 2023).
8. Life Cycle Analysis - an overview | ScienceDirect Topics, https://www.sciencedirect.com/topics/earth-and-planetary-sciences/life-cycle-analysis, (accessed July 20, 2023).
9. B. Steen, *Management of Environmental Quality: An International Journal*, 2005, **16**, 107–118.
10. Environmental and economic assessment of waste collection and transportation using LCA: A case study - ScienceDirect, https://www-sciencedirect-com.libproxy1.nus.edu.sg/science/article/pii/S0013935123009003, (accessed July 20, 2023).
11. E. Sai, R. Koide and S. Murakami, *Sustainable Production and Consumption*, 2023, **38**, 331–340.
12. T. Schaubroeck, *Frontiers in Sustainability*, DOI: https://doi.org/10.3389/frsus.2023.1063583.
13. Prospective consequential Life Cycle Assessment: Identifying the future marginal suppliers using Integrated Assessment Models | Energy | ChemRxiv | Cambridge Open Engage, https://chemrxiv.org/engage/chemrxiv/article-details/63ee10cdfcfb27a31fe227df, (accessed July 20, 2023).
14. Why and when?, https://consequential-lca.org/clca/why-and-when/, (accessed July 20, 2023).
15. Why and when?, https://consequential-lca.org/clca/why-and-when/, (accessed December 15, 2023).
16. F. Creutzig, A. Popp, R. Plevin, G. Luderer, J. Minx and O. Edenhofer, *Nature Clim Change*, 2012, **2**, 320–327.
17. Life Cycle Assessment (LCA) explained, https://pre-sustainability.com/articles/life-cycle-assessment-lca-basics/, (accessed December 15, 2023).
18. L. Nickel, How to define the Goal & Scope of your LCA, https://ecochain.com/knowledge/how-to-define-the-goal-scope-of-your-lca/, (accessed July 20, 2023).
19. J. Navarro and F. Zhao, *Frontiers in Energy Research*.
20. M. A. Curran, in *Goal and Scope Definition in Life Cycle Assessment*, ed. M. A. Curran, Springer Netherlands, Dordrecht, 2017, pp. 1–62.
21. Life Cycle Assessment (LCA) explained, https://pre-sustainability.com/articles/life-cycle-assessment-lca-basics/, (accessed July 20, 2023).
22. Goal and Scope Definition in Life Cycle Assessment, https://www.springerprofessional.de/en/goal-and-scope-definition-in-life-cycle-assessment/10736652, (accessed July 20, 2023).
23. A. Bjørn, A. Moltesen, A. Laurent, M. Owsianiak, A. Corona, M. Birkved and M. Z. Hauschild, in *Life Cycle Assessment: Theory and Practice*, eds. M. Z. Hauschild, R. K. Rosenbaum and S. I. Olsen, Springer International Publishing, Cham, 2018, pp. 117–165.
24. T. Ekvall and B. P. Weidema, *Int. J. LCA*, 2004, **9**, 161–171.
25. S. Suh and G. Huppes, *Journal of Cleaner Production*, 2005, **13**, 687–697.
26. openLCA Nexus: The source for LCA data sets, https://nexus.openlca.org/, (accessed July 20, 2023).
27. R. H. Crawford, P.-A. Bontinck, A. Stephan, T. Wiedmann and M. Yu, *Journal of Cleaner Production*, 2018, **172**, 1273–1288.

28. C. Spreafico, D. Landi and D. Russo, *Sustainable Production and Consumption*, 2023, **38**, 241–251.
29. J. Nakatani, *Sustainability*, 2014, **6**, 6158–6169.
30. Y. Yang and R. Heijungs, *Int J Life Cycle Assess*, 2018, **23**, 751–758.
31. T. Xayachak, N. Haque, D. Lau, R. Parthasarathy and B. K. Pramanik, *Process Safety and Environmental Protection*, 2023, **173**, 592–603.
32. J. Li, Y. Tian and K. Xie, *Ecological Indicators*, 2023, **153**, 110455.
33. A. Azapagic and R. Clift, *Int. J. LCA*, 1998, **3**, 305–316.
34. G. Wernet, C. Bauer, B. Steubing, J. Reinhard, E. Moreno-Ruiz and B. Weidema, *Int J Life Cycle Assess*, 2016, **21**, 1218–1230.
35. L. Zakrisson, E. S. Azzi and C. Sundberg, *Int J Life Cycle Assess*, 2023, **28**, 907–923.
36. G. Mai, W. Huang, J. Sun, S. Song, D. Mishra, N. Liu, S. Gao, T. Liu, G. Cong, Y. Hu, C. Cundy, Z. Li, R. Zhu and N. Lao, On the Opportunities and Challenges of Foundation Models for Geospatial Artificial Intelligence, https://arxiv.org/abs/2304.06798v1, (accessed July 20, 2023).
37. G. Giusti, D. V. da Silva, A. C. G. Albino, Y. de Souza Tadano and D. A. L. Silva, *Int J Life Cycle Assess*, DOI: https://doi.org/10.1007/s11367-023-02184-8.
38. M. R. Seyedabadi, S. Samareh Abolhassani and U. Eicker, *Journal of Building Engineering*, 2023, **76**, 107101.
39. L. Riondet, M. Rio, V. Perrot-Bernardet and P. Zwolinski, *Procedia CIRP*, 2023, **116**, 714–719.
40. S. Langkau, B. Steubing, C. Mutel, M. P. Ajie, L. Erdmann, A. Voglhuber-Slavinsky and M. Janssen, *Int J Life Cycle Assess*, DOI: https://doi.org/10.1007/s11367-023-02175-9.
41. R. N. Hansen, F. N. Rasmussen, M. Ryberg and H. Birgisdóttir, *Int J Life Cycle Assess*, 2023, **28**, 131–145.
42. S. Bruhn, R. Sacchi, C. Cimpan and M. Birkved, *Building and Environment*, 2023, **242**, 110535.
43. G. Guignone, J. L. Calmon, D. Vieira and A. Bravo, *Journal of Building Engineering*, 2023, **73**, 106780.
44. V. De Laurentiis, A. Amadei, E. Sanyé-Mengual and S. Sala, *Int J Life Cycle Assess*, , DOI: https://doi.org/10.1007/s11367-023-02188-4.
45. N. Quernheim, S. Winter, L. Arnemann and B. Schleich, *Proceedings of the Design Society*, 2023, **3**, 2635–2644.
46. S. Minner and M. Yao, *Review of Multi-Supplier Inventory Models in Supply Chain Management: An Update*, 2017.
47. A. Federgruen and P. Zipkin, *Mathematics of Operations Research*, 1986, **11**, 193–207.
48. C. Ugaya, J. B. de Araújo, A. Souza, B. B. T. do Carmo, S. A. de Oliveira and V. G. Maciel, *Int J Life Cycle Assess*, 2023, **28**, 199–218.
49. R. K. Rosenbaum, M. Z. Hauschild, A.-M. Boulay, P. Fantke, A. Laurent, M. Núñez and M. Vieira, in *Life Cycle Assessment: Theory and Practice*, eds. M. Z. Hauschild, R. K. Rosenbaum and S. I. Olsen, Springer International Publishing, Cham, 2018, pp. 167–270.
50. J. C. Bare, *Clean Techn Environ Policy*, 2010, **12**, 341–351.
51. M. Z. Hauschild, M. Goedkoop, J. Guinée, R. Heijungs, M. Huijbregts, O. Jolliet, M. Margni, A. De Schryver, S. Humbert, A. Laurent, S. Sala and R. Pant, *Int J Life Cycle Assess*, 2013, **18**, 683–697.
52. S. V. Withanage and K. Habib, *Sustainability*, 2021, **13**, 7939.
53. M. A. Curran, *Life Cycle Assessment Handbook: A Guide for Environmentally Sustainable Products*, John Wiley & Sons, 2012.
54. D. Laner and H. Rechberger, in *Special Types of Life Cycle Assessment*, ed. M. Finkbeiner, Springer Netherlands, Dordrecht, 2016, pp. 293–332.
55. M. Damiani, T. Sinkko, C. Caldeira, D. Tosches, M. Robuchon and S. Sala, *Environmental Impact Assessment Review*, 2023, **101**, 107134.
56. J. Ferdous, F. Bensebaa and N. Pelletier, *Journal of Cleaner Production*, 2023, **402**, 136804.
57. H. Mostafaei, Z. Keshavarz, M. A. Rostampour, D. Mostofinejad and C. Wu, *Structures*, 2023, **53**, 279–295.

58. M. Smurthwaite, L. Jiang and K. S. Williams, *Environ Dev Sustain*, DOI:https://doi.org/10.1007/s10668-023-03453-0.
59. M. Niero and P. P. Kalbar, *Resources, Conservation and Recycling*, 2019, **140**, 305–312.

Chapter 3
Sustainable Strategies for Oil and Gas and Steel Industries

Abstract This chapter explores the growing importance of sustainability in the oil and gas industry and the steel industry. It provides background information on both industries, highlighting their significance and impact. The study examines the environmental impact and climate change concerns, resource depletion and conservation issues, and the social responsibility and stakeholder expectations faced by these industries. Current initiatives and barriers to sustainability implementation are discussed, followed by an analysis of the industrial scenario. Strategies to implement sustainability practices are presented, covering various stages of the oil and gas production process, including exploration, drilling fluids, well completion, production, and surface processing. Additionally, strategies specific to steel manufacturing are outlined. The article also addresses the importance of sustainable reporting for these industries. Finally, the conclusion summarizes the key findings and highlights the need for continued efforts towards sustainability in the oil and gas and steel sectors.

3.1 Introduction

3.1.1 Background of the Oil and Gas Industry

The oil and gas industry has a rich and complex background that spans several decades. It has played a pivotal role in shaping global economies, driving technological advancements, and meeting the world's energy demands. To understand the significance of this industry, it is essential to explore its historical development, production levels, and economic impact. The roots of the modern oil and gas industry can be traced back to the mid-nineteenth century when the first commercial oil well was drilled in Titusville, Pennsylvania, in 1859. This marked the beginning of a transformative era as the world discovered the potential of oil as a source of energy. The subsequent years witnessed the rapid expansion of oil exploration and production, primarily driven by the industrialization and urbanization of societies. Figure 3.1 illustrates the evolutionary history of the global natural gas industry. One of the most

S. Ramakrishna and B. Ramasubramanian, *Handbook of Materials Circular Economy*,
https://doi.org/10.1007/978-981-97-0589-4_3

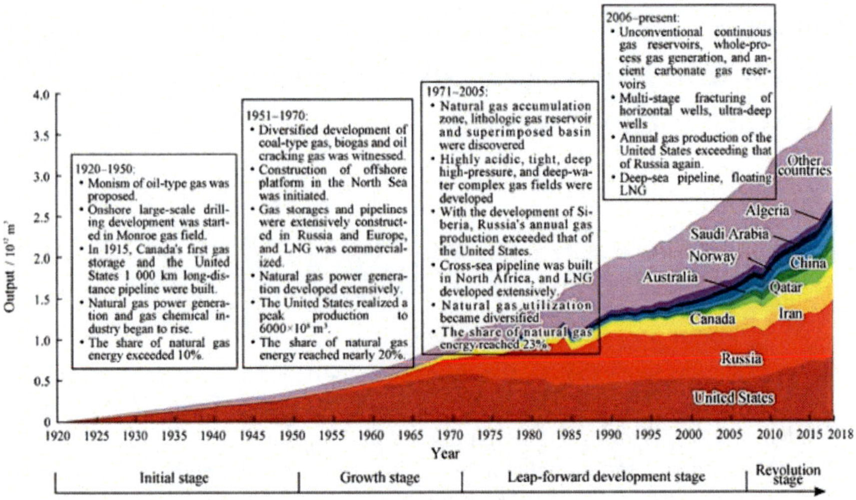

Fig. 3.1 Evolution history of the world natural gas industry [5]

significant milestones in the industry's history was the discovery of vast oil reserves in the Middle East during the early twentieth century. The region, particularly countries like Saudi Arabia, Iran, Iraq, and Kuwait, emerged as major players in the global oil market, possessing significant oil reserves that still hold importance today. The formation of the Organization of the Petroleum Exporting Countries (OPEC) in 1960 further consolidated their influence and allowed for greater control over oil prices and production levels [1].

The global oil and gas industry has witnessed exponential growth over the years. In terms of oil production, the top oil-producing countries include the United States, Saudi Arabia, Russia, and China. According to data from the International Energy Agency (IEA), global oil production reached approximately 100 million barrels per day (bpd) in 2021, with the United States being the largest producer, accounting for around 17% of the total production. In terms of proven oil reserves, Saudi Arabia, Venezuela, Canada, and Iran hold significant portions. Saudi Arabia alone possesses around 17% of the world's proven oil reserves. These reserves are essential for ensuring long-term energy security and meeting the growing global energy demand. The natural gas sector has also seen substantial growth and development. Natural gas is increasingly being recognized as a cleaner alternative to coal and oil, contributing to lower carbon emissions. The top natural gas-producing countries include the United States, Russia, Iran, and Qatar. The United States has experienced a shale gas revolution, thanks to advancements in hydraulic fracturing technology, making it the world's leading natural gas producer [2].

In terms of consumption, the oil and gas industry dominate the global energy landscape. According to the IEA, oil accounted for around 33% of the world's total energy consumption in 2020, while natural gas accounted for approximately 24%. These figures highlight the industry's crucial role in meeting the energy demands of various sectors, including transportation, manufacturing, and residential use. Economically, the oil and gas industry has a significant impact on global economies. It attracts substantial investments in exploration, production, refining, and distribution infrastructure. According to the World Bank, the industry's contribution to global GDP was approximately 2.8% in 2020. Additionally, the industry provides employment opportunities and contributes to government revenues through taxes, royalties, and other fees [3, 4].

However, it is important to note that the oil and gas industry also faces challenges and uncertainties. The fluctuating oil prices, geopolitical tensions, environmental concerns, and the need to transition to renewable energy sources pose significant challenges to the industry's long-term sustainability. The growing global focus on reducing carbon emissions and mitigating climate change is pushing the industry to adopt cleaner technologies and invest in renewable energy solutions [6]. Over the years, the oil and gas industry has undergone remarkable transformations, with technological advancements enabling the extraction of hydrocarbons from increasingly challenging environments. This has led to the discovery and exploitation of vast reserves in offshore areas, deepwater locations, and unconventional resources like shale gas and oil sands. These developments have propelled the industry into a position of immense significance, not only as a source of energy but also as a driver of economic growth and geopolitical influence [7].

3.1.2 Background of the Steel Industry

The origins of steelmaking can be traced back to ancient times, with early civilizations utilizing iron and its alloys for tools, weapons, and construction. However, it was during the Industrial Revolution in the eighteenth and nineteenth centuries that the steel industry witnessed significant advancements. Innovations in production processes, such as the Bessemer process and the open-hearth furnace, revolutionized steel manufacturing and fueled the rapid growth of industries such as railways, shipbuilding, and construction. Today, the steel industry remains a vital component of global economies, supporting various sectors and infrastructure development. In terms of production, China has been the world's largest steel producer for several years. According to data from the World Steel Association, China accounted for around 58% of global crude steel production in 2020, producing approximately 1.05 billion metric tons of crude steel. Other significant steel-producing countries include India, Japan, the United States, and Russia [8]. Figure 3.2 illustrates the trend or pattern of steel production on a global scale from 1965 to 2020.

In terms of consumption, the construction and infrastructure sectors are the largest steel consumers globally. Steel is a fundamental material in the construction of buildings, bridges, roads, and other infrastructure projects. The automotive and manufacturing sectors are also major consumers of steel, using it for the production of vehicles, machinery, appliances, and various consumer goods. The steel industry's economic impact is significant, contributing to employment, GDP, and trade. According to the

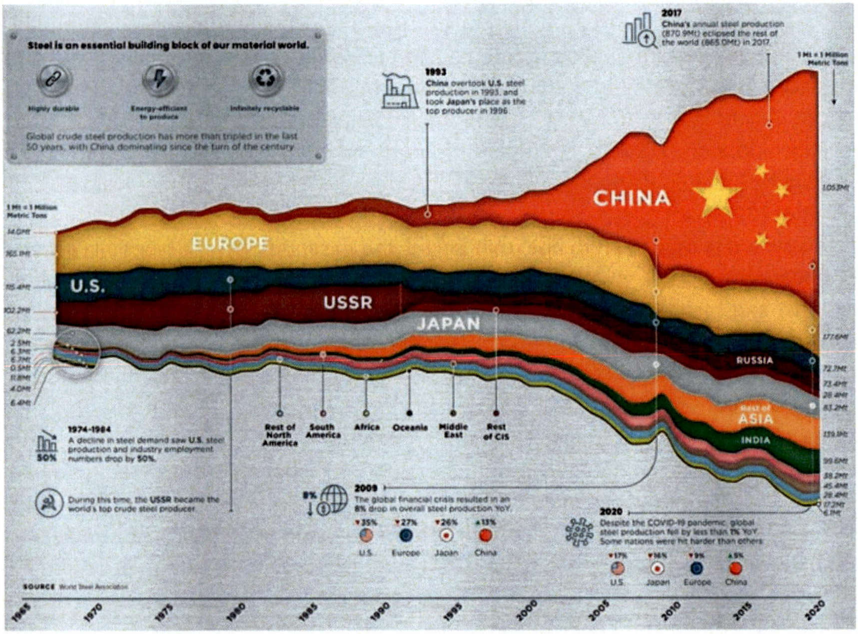

Fig. 3.2 Global trend of Steel production [9]

World Steel Association, the global steel industry contributed approximately 2.6% to the world's GDP in 2020. It supports millions of direct and indirect jobs worldwide, with steelmaking and related activities serving as a source of livelihood for many communities. The industry also generates substantial trade, with steel products being traded globally to meet demand in different regions [10, 11].

In terms of steel production technologies, the industry has evolved to become more efficient and environmentally friendly. Traditional blast furnaces, which rely on the combustion of coke and coal, still dominate steel production. However, there is a growing shift towards electric arc furnaces (EAFs) that use scrap metal as a raw material and consume less energy [12]. EAFs also contribute to recycling and circular economy efforts by utilizing scrap steel and reducing the reliance on virgin iron ore. The steel industry faces several challenges and opportunities in the current global landscape. One of the key challenges is the environmental impact of steel production. The steel industry is a significant emitter of greenhouse gases, primarily carbon dioxide (CO_2), due to the combustion of fossil fuels and the chemical reactions involved in the production processes. Efforts are underway to develop cleaner technologies, improve energy efficiency, and reduce carbon emissions in steelmaking, including the exploration of hydrogen-based processes and carbon capture and storage (CCS) solutions. Moreover, the steel industry faces increasing pressure to adopt sustainable practices and reduce its ecological footprint. This includes responsible sourcing of raw materials, managing waste and by-products, and implementing efficient water and energy management systems [13, 14]. Global steel demand and required decrease of CO_2 emissions in 2050 and historical data of world steel production has been depicted respectively in Fig. 3.3a, b.

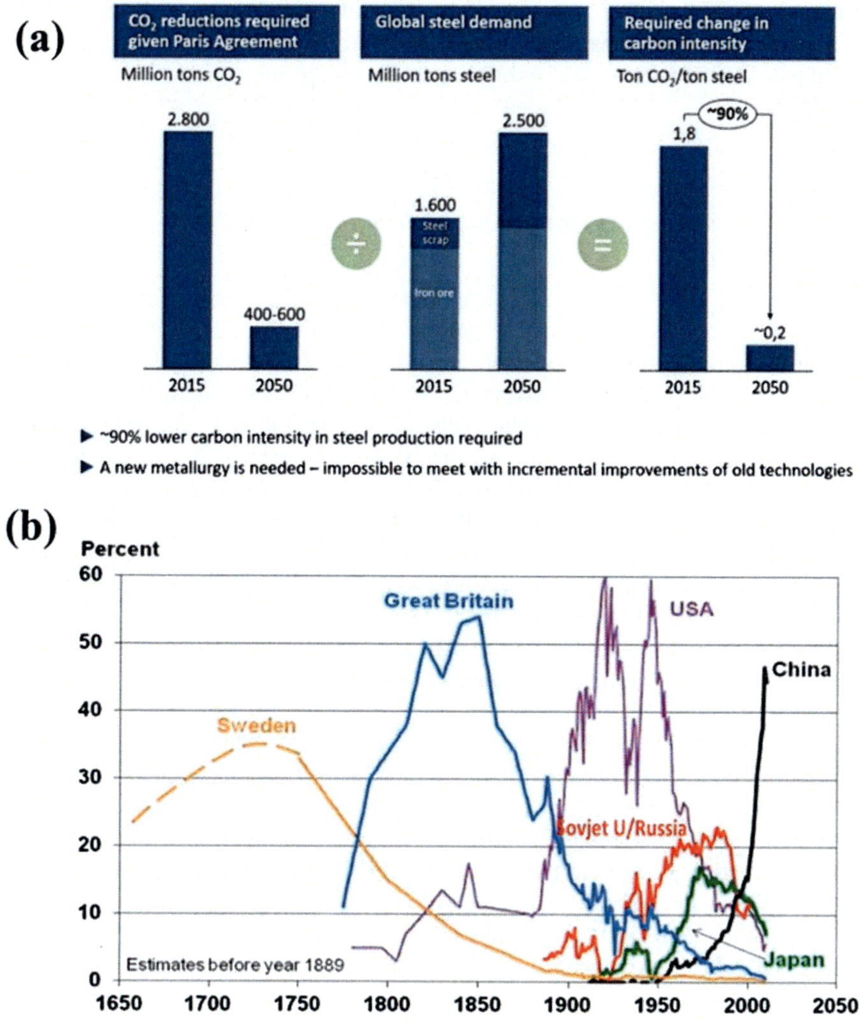

Fig. 3.3 **a** Projected global steel demand and CO_2 emission reduction targets for 2050 [15]. **b** Historical data of world steel production [15]

3.2 Growing Importance of Sustainability in the Oil and Gas and Steel Industries

In recent years, there has been an increasing recognition of the urgent need for sustainability across various industries, including oil and gas and steel. The growing concerns about climate change, environmental degradation, and the depletion of natural resources have prompted a shift in mindset and practices within these sectors [16, 17].

3.2.1 Environmental Impact and Climate Change

The oil and gas industry has a significant environmental footprint, contributing to various adverse effects. The extraction and processing of hydrocarbons give rise to air and water pollution, habitat destruction, and the release of greenhouse gases (GHGs) like carbon dioxide (CO2) and methane (CH4). According to the International Energy Agency (IEA), the oil and gas sector was responsible for approximately 35% of global CO_2 emissions in 2020 (International Energy Agency (IEA). (2020). Global Energy Review 2020). Methane, a potent GHG, is emitted during different activities in the industry, including drilling, production, and transportation. It is estimated that the oil and gas sector accounted for around 25% of global methane emissions (Intergovernmental Panel on Climate Change (IPCC). (2019). IPCC Special Report on Climate Change and Land) [18, 19].

Similarly, the steel industry contributes significantly to GHG emissions. The production processes involve the combustion of fossil fuels, predominantly coal and coke, resulting in the release of CO_2. In 2020, the global steel sector accounted for approximately 7% of global CO_2 emissions (World Steel Association. (2020). Steel's Contribution to a Low Carbon Future and Climate Resilient Societies). Recognizing the need for emission reductions, the industry is exploring cleaner technologies such as hydrogen-based processes and carbon capture and storage (CCS) (International Energy Agency (IEA). (2021). Energy Technology Perspectives 2020). However, the large-scale implementation of these technologies poses challenges and requires further development [20]. Fig. 3.4a presents a comprehensive global overview of carbon pricing, emissions trading schemes, and other initiatives. Figure 3.4b illustrates the carbon impact on steel.

3.2.2 Resource Depletion and Conservation

Both the oil and gas and steel industries face challenges related to resource depletion. The finite nature of oil and gas reserves raises concerns about energy security and long-term availability. According to the BP Statistical Review of World Energy,

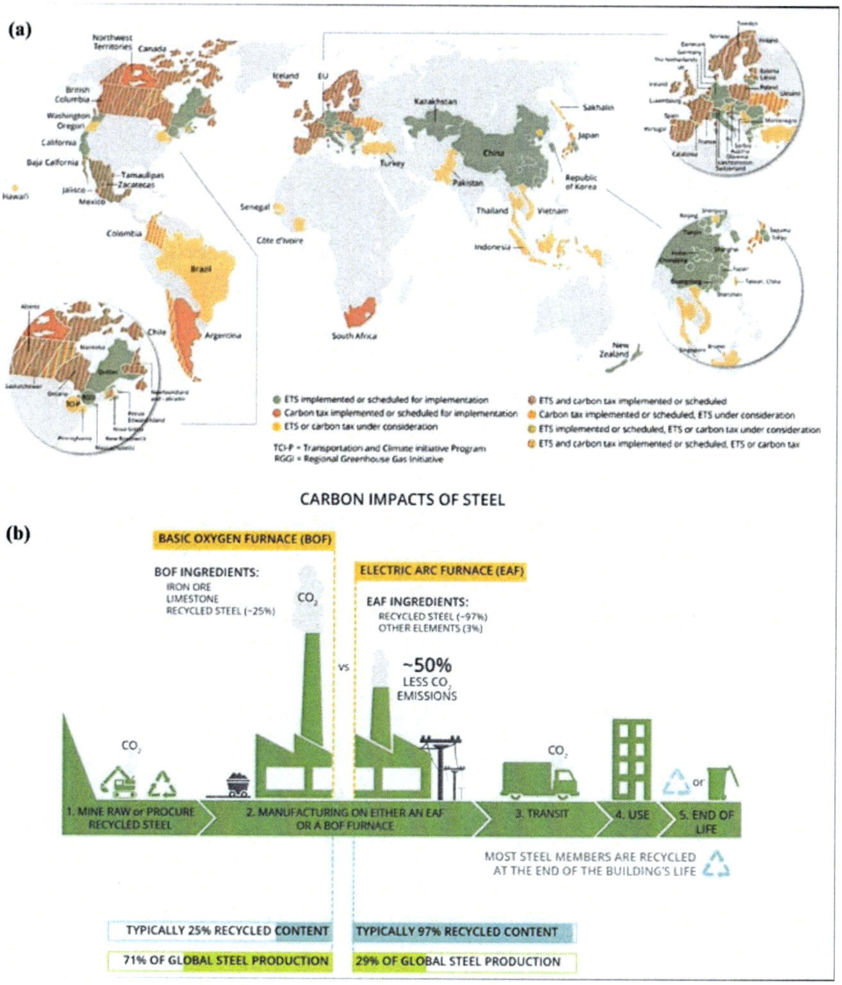

Fig. 3.4 **a** Global overview of carbon pricing, emissions trading schemes, and other initiatives [21]. **b** Carbon impacts on steel [22]

global proven oil reserves stood at approximately 1.7 trillion barrels in 2020 (BP. (2021). BP Statistical Review of World Energy 2021). However, the rate of new reserve discoveries has been declining over the years, underscoring the importance of diversifying the energy mix and exploring renewable alternatives [23]. Figure 3.5 displays statistical data regarding the primary CO_2 emission point-sources within the coal industry, spanning from resource extraction to end-use.

The steel industry heavily relies on finite resources such as iron ore and coal. The extraction and processing of these raw materials have significant environmental impacts. For instance, iron ore mining often involves deforestation and habitat destruction, disrupting ecosystems. Efforts are being made to optimize resource

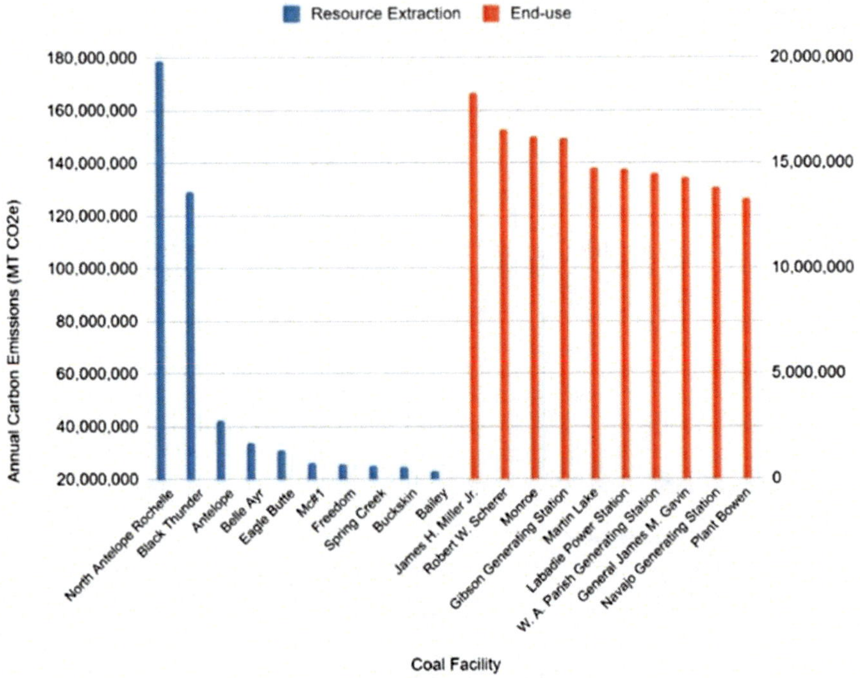

Fig. 3.5 Primary CO_2 emission point-sources in the coal industry: from resource extraction to end-use [24]

utilization and reduce reliance on virgin raw materials. The global steel recycling rate reached approximately 88% in 2020, demonstrating progress in reducing demand for new resources (World Steel Association. (2020). Sustainability Indicators 2020) [25].

3.2.3 Social Responsibility and Stakeholder Expectations

Social responsibility and meeting stakeholder expectations are critical considerations for both the oil and gas and steel industries. Communities residing near oil and gas operations often experience adverse effects, including noise pollution, land disturbance, and potential health hazards. Engaging with local communities, respecting indigenous rights, and implementing robust environmental and social impact assessments are essential for sustainable operations. The steel industry also faces challenges related to worker safety, labor rights, and responsible supply chain management. Stakeholders, including employees, customers, investors, and civil society, demand increased transparency, ethical practices, and responsible governance. Companies are actively improving worker safety measures, ensuring fair labour practices, and implementing sustainable supply chain management systems [26].

Note

Navigating the transition to a new energy landscape requires concerted efforts and strategic actions. McKinsey and Company (2022a) have identified three key actions that play a crucial role in this process:

(a) Creating an enabling environment to foster investments in renewable energy projects. This involves establishing supportive policies, regulatory frameworks, and incentives that encourage the development and deployment of renewable energy technologies. By creating a conducive ecosystem, governments and industry stakeholders can attract investments and drive the transition towards a sustainable energy future.
(b) Enhancing accessibility to capital pools, financing, and investments specifically targeted at renewable projects. Access to finance is often cited as a major barrier to the widespread adoption of renewable energy solutions. By developing innovative financial mechanisms, such as green bonds, venture capital funds, and public–private partnerships, barriers to financing can be overcome, unlocking the potential for renewable energy expansion.
(c) Strengthening the capabilities of the local workforce in the oil and gas industry to support the growth of sustainable energy businesses. As the energy landscape evolves, there is a need to equip the workforce with the necessary skills and knowledge to operate in a sustainable energy ecosystem. This can be achieved through training programs, capacity building initiatives, and collaboration with educational institutions to ensure a skilled workforce that can drive the development and deployment of sustainable energy solutions.

Aramco's report (2021) exemplifies the implementation of circularity principles within their operations. The company has embraced innovative approaches, such as the use of modular, skid-mounted structures in the Midyan gas plant, resulting in substantial cost savings. Additionally, their scrap-to-commodity program has yielded over $30 million by recycling materials and reintroducing them as feedstocks to local manufacturers. Aramco's commitment to conserving water resources is evident through the reuse of wastewater for irrigation and cooling purposes, effectively reducing groundwater consumption [27].

Reliance Industries (2020) prioritizes the diligent utilization of scarce resources. Through various measures, including recycling initiatives and converting organic waste into valuable resources like manure and biogas, the company improves raw material productivity while minimizing waste. Additionally, their logistics operations focus on using bulk tankers, reducing packaging materials, handling, and containment, which not only benefits the company but also contributes to sustainability throughout the supply chain [28].

Shell's sustainability report (2021) underscores the significance of improved circularity in the global plastic market. The company actively promotes the reduction, reuse, and recycling of plastics, aiming to minimize waste and environmental impact. In line with their commitment to circularity, Shell has developed a low-temperature bitumen solution for road surfaces, reducing waste and enhancing sustainability in the construction sector [29].

Ongcindia (2020) recognizes the importance of carbon capture, utilization, and storage (CCUS) technologies in mitigating CO_2 emissions. These technologies offer opportunities to decarbonize major industrial sectors, including steel, cement, paper, refining, and petrochemicals. Ongcindia's adoption of GHG accounting and mitigation projects enables a comprehensive assessment of emissions and paves the way for targeted reduction measures. Additionally, the organization proactively addresses potential water crises by implementing sustainable water management strategies, including reuse, reduction, recycling, and replenishment, while encouraging the use of recycled water in their operations [30].

3.3 Current Initiatives and Barriers

Methodology: To gain insights into the progress and challenges surrounding the integration of sustainability practices in the oil and gas (O&G) industry, a comprehensive literature review was conducted using the Scopus database. The search strategy included the following search terms:

(TITLE-ABS-KEY ("Sustainability" AND challenges AND "oil and gas").
TITLE-ABS-KEY ("Sustainable practices" AND barriers AND "petroleum industry").
TITLE-ABS-KEY ("Environmental impact" AND constraints AND "energy sector").
TITLE-ABS-KEY ("Renewable energy" AND limitations AND "oil and gas production").
TITLE-ABS-KEY ("Carbon emissions" AND challenges AND "hydrocarbon industry").

After filtering out irrelevant papers, non-English publications, and conference proceedings, a total of 140 articles remained for further analysis. To identify the most pertinent papers, the abstracts of these articles were carefully reviewed in relation to the research questions, resulting in the selection of 33 articles that provided valuable insights into 4 important sustainable concepts in the O&G industry as listed in the below section,

Countries worldwide have taken significant strides towards sustainability in the oil and gas and steel industries. Norway pioneers this movement, implementing carbon pricing, stringent regulations, and investing in clean technologies. The United Arab Emirates focuses on renewable energy investments, reduced gas flaring, and enhanced water conservation. Saudi Arabia promotes economic diversification through renewable projects and energy efficiency measures. Germany leads in steel sustainability, setting emission targets and emphasizing circular economy principles. China enforces environmental regulations and prioritizes energy efficiency and circular practices. Japan emphasizes resource conservation and renewable energy adoption in both sectors. Current initiatives in the oil and gas and steel industries with regards to sustainability encompass a range of practices and programs aimed at reducing environmental impact, improving resource efficiency, and promoting responsible operations [31]. Here are some notable initiatives:

(i) Carbon Capture, Utilization, and Storage (CCUS) is a technology aimed at capturing carbon dioxide (CO_2) emissions from industrial processes, including oil and gas operations, and storing them to prevent their release into the atmosphere. CO_2 is captured from flue gases or industrial emissions using different capture methods such as post-combustion capture, pre-combustion capture, or oxyfuel combustion. Each method has its own set of parameters and technologies to optimize the capture efficiency and minimize energy consumption [32]. Once captured, the CO_2 needs to be transported from the capture site to the storage site. This often involves compression and the use of pipelines or ships to transport the CO_2 over long distances. Parameters such as pressure, temperature, and flow rate need to be carefully controlled to ensure safe and efficient transportation. The captured CO_2 is stored deep underground in geological formations, typically in depleted oil and gas reservoirs, saline aquifers, or unmineable coal seams. The storage process involves injecting the CO_2 into these formations and monitoring its behavior to ensure it remains securely trapped and isolated from the atmosphere. Parameters like injection rate, pressure, and monitoring techniques play a crucial role in the effectiveness and safety of CO_2 storage. Norway has been a leader in CCUS technology and has established the Northern Lights project, which aims to develop infrastructure for large-scale CO_2 capture, transport, and storage in the North Sea. The United States has various CCUS projects, including the Petra Nova project in Texas, which captures CO_2 from a coal-fired power plant and stores it in an underground reservoir. Canada has the Quest project in Alberta, which captures CO_2 from an oil sands upgrader and stores it underground. The UK has the Acorn project in Scotland, which aims to establish a carbon capture and storage hub [33, 34].

Technical difficulties in CCUS implementation include the high costs associated with capture, transportation, and storage infrastructure, as well as the energy requirements for the capture process. Ensuring the long-term integrity and security of stored CO_2, addressing public acceptance and regulatory challenges, and scaling up CCUS technologies to achieve significant emission reductions are additional complexities to be addressed [35].

(ii) Renewable energy integration in oil and gas operations involves installing solar panels to convert solar radiation into electricity and erecting wind turbines to capture wind energy. Energy storage systems, such as batteries or compressed air storage, are used to store excess renewable energy. Parameters considered include capacity, output, grid integration, and power management systems. Country initiatives include the UAE's investments in solar energy and Norway's integration of hydropower. Technical challenges include managing intermittency, addressing grid stability issues, developing infrastructure, and establishing supportive regulatory frameworks [36].

(iii) Methane emission reduction in the O&G industry involves the use of advanced monitoring and leak detection systems, best practices in drilling and production, and various process parameters. Monitoring frequency and leak detection sensitivity are key considerations, along with robust data analysis and reporting systems. Country initiatives include the Methane Challenge Program in the United States and regulatory measures in Canada. Technical challenges include the detection and localization of leaks, maintenance and upkeep of infrastructure, measurement accuracy, and cost and implementation issues.

(iv) Water management and conservation strategies in the O&G industry reduce freshwater consumption by utilizing alternative sources like treated wastewater and brackish water for non-potable purposes. A crucial aspect is treating and reusing produced water extracted during drilling. Treated produced water finds application in hydraulic fracturing and secondary recovery operations, enhancing oil and gas extraction. Filtration removes solid particles, sediments, and suspended materials. Membrane processes separate contaminants at a molecular level, eliminating dissolved salts and organic compounds. Biological treatment employs microorganisms to purify water by consuming or converting pollutants. These methods effectively conserve freshwater resources while promoting sustainable practices [37, 38].

3.4 Industrial Scenario

It is evident that many industries share common goals in addressing climate change and transitioning to a more sustainable future. Companies like ExxonMobil, BP, Chevron, and TotalEnergies have set ambitious targets to reduce emissions and improve energy efficiency. For instance, ExxonMobil aims to reduce methane emissions by 15% and flaring intensity by 25% by 2025, while Chevron invests in renewable energy projects and supports research on carbon capture and storage. BP is committed to becoming a net-zero emissions company by 2050 and invests in renewable energy and biofuels. TotalEnergies focuses on expanding renewable energy capacity and reducing the carbon intensity of its operations. These companies demonstrate a shift towards a more sustainable energy portfolio [39].

Furthermore, while the top oil and gas companies contribute significantly to global energy production, their statistics reveal varying approaches to sustainability (Table 3.1). Some companies prioritize renewable energy investments. For example, Equinor focuses on offshore wind projects, while TotalEnergies commits to reducing flaring and advancing biofuels. Others concentrate on improving energy efficiency and reducing emissions. PetroChina implements measures to reduce emissions intensity, and Occidental Petroleum focuses on carbon capture and storage projects. However, there are also companies that face challenges in their sustainability journey. For instance, Gazprom emphasizes natural gas as a lower-carbon alternative, but its progress in renewable energy integration is relatively limited. The statistics and initiatives of these companies reflect the diverse strategies adopted within the industry to address sustainability goals [40].

IPIECA stands for the International Petroleum Industry Environmental Conservation Association [41]. During COP27, IPIECA actively participated in negotiations and hosted events to explore pathways for a sustainable energy transition. They organized side events on business engagement, net-zero emissions, and the methane challenge. IPIECA aims to collaborate with stakeholders for a just transition in the oil and gas industry. They will share outcomes with members to align their workstreams with the Paris Agreement. Similar organizations include WBCSD, Global Compact Network, and ICC, all engaging in sustainable development and corporate responsibility.

3.5 Strategies to Implement

The oil and gas industry is undergoing an intriguing transformation, shifting its focus from regulatory compliance to a more comprehensive approach that prioritizes sustainability. This shift emphasizes the industry's commitment to actively reduce emissions and support the energy transition while recognizing the importance of health, safety, and environmental considerations. The range of stakeholders involved has expanded significantly, now encompassing not only shareholders, regulators, employees, and local communities but also customers, investors, non-governmental institutions, and society as a whole. Consequently, reporting practices have evolved from being voluntary and sporadic to becoming standardized, regular, and transparent, ensuring greater accountability. In terms of governance, there has been a notable change within the industry. Sustainability oversight and accountability, previously embedded within health, safety, and environment functions, have now been elevated to board-level involvement. This shift signifies that sustainability is no longer viewed as a secondary aspect of corporate citizenship but rather a critical pillar for long-term business competitiveness [42, 43].

To effectively reduce emissions in oil and gas operations, companies are required to take several actions. This includes mapping energy flows and emissions, benchmarking performance internally and externally, setting ambitious targets, and implementing regular reporting. Additionally, companies must pilot and deploy advanced technologies that can effectively monitor and mitigate emissions, while also developing new enabling technologies for emission mitigation. They need to adapt their investment screening criteria to consider the anticipated impact of future regulations and manage their portfolios accordingly. Strengthening internal governance and performance management structures is crucial to effectively support the achievement of sustainability objectives. It is important to acknowledge that while all oil and gas companies are expected to reduce emissions, not all will transition into diversified energy companies. To determine the role of low carbon energy in their future portfolios, companies must address key questions regarding alignment with their

Table 3.1 Companies and their initiatives

Company	Sustainability initiatives	Additional details
ExxonMobil	Invests in low-carbon technologies and carbon capture utilization and storage (CCUS) projects	
Saudi Aramco	Focuses on carbon emissions reduction through energy efficiency programs and research on low-carbon technologies	
Royal Dutch Shell	Invests in biofuels, electric vehicle charging infrastructure, and renewable energy sources	
British Petroleum	Aims to become a net-zero emissions company by 2050. Invests in renewable energy, electric vehicle charging, and biofuels	
Chevron	Invests in renewable energy projects, including wind and solar, and supports research on carbon capture, utilization, and storage	
Total Energies	Focuses on expanding renewable energy capacity, investing in battery storage, and reducing the carbon intensity of its operations	
Gazprom	Invests in energy efficiency measures and expands the use of natural gas as a lower-carbon alternative to coal and oil	
PetroChina	Invests in natural gas exploration and production, as well as carbon capture, utilization, and storage projects	
Equinor	Focuses on renewable energy investments, including offshore wind, and aims to reduce the carbon intensity of its operations	
ConocoPhillips	Sets emissions reduction targets and invests in research and development of low-carbon technologies	
Eni	Invests in renewable energy projects, carbon capture, utilization, and storage, and aims to reduce carbon intensity and methane emissions	
Rosneft	Implements energy efficiency measures and invests in new technologies to reduce greenhouse gas emissions	
Occidental Petroleum	Focuses on carbon capture, utilization, and storage projects and aims to reduce emissions intensity	
1Kuwait Petroleum Corporation	Invests in energy efficiency measures and explores opportunities for renewable energy integration	
Sinopec	Invests in low-carbon technologies and energy efficiency measures	
Repsol	Aims to achieve net-zero emissions by 2050 and invests in renewable energy projects and circular economy initiatives	
Lukoil	Implements energy efficiency measures and invests in reducing emissions intensity	

(continued)

Table 3.1 (continued)

Company	Sustainability initiatives	Additional details
Petrobras	Focuses on reducing emissions intensity, improving energy efficiency, and investing in renewable energy projects	
Abu Dhabi National Oil Company (ADNOC)	Invests in carbon capture, utilization, and storage projects and aims to reduce carbon intensity	
Marathon Petroleum	Focuses on energy efficiency measures and invests in renewable energy projects	
Hess Corporation	Aims to reduce emissions intensity and invests in renewable energy and carbon capture projects	
Emirates National Oil Corporation (ENOC)	Implements energy efficiency measures and invests in renewable energy projects	
Petroleum Authority of Thailand (PTT)	Invests in renewable energy projects and energy efficiency measures	
Reliance Industries	Focuses on energy efficiency measures and invests in renewable energy projects	

existing capabilities and technical expertise [44]. They need to consider the potential for development and exploitation within the low carbon energy sector. Financial resources required for successful diversification need to be carefully considered, striking a balance between investments in low carbon energy and core oil and gas operations. Additionally, companies must evaluate how their operating models should evolve to accommodate new business sectors. This includes exploring options such as integrating new ventures to leverage shared services and synergies or managing them at arm's length. Furthermore, the balance between returns from oil and gas and future growth in low carbon must be considered to define the integrated value proposition for the company [45].

3.5.1 Exploration

Exploration in the oil and gas industry encompasses various practices aimed at identifying potential reservoirs. Geologists conduct geological surveys by analyzing surface rocks, sedimentary formations, and geological structures to ascertain areas with hydrocarbon accumulation potential. By examining composition, age, and structure, they determine the likelihood of finding oil and gas reserves. Geophysicists employ techniques like seismic, gravity, magnetic, and electromagnetic surveys to gather subsurface data, identifying features and anomalies indicating the presence of reservoirs. Subsequently, collected geological and geophysical data undergoes processing and analysis, utilizing advanced software and algorithms to generate models and maps of subsurface structures, aiding in the identification of potential reservoirs [46].

The effectiveness of exploration relies on the utilization of various tools and technologies for data gathering and analysis. Seismic surveying involves generating and recording sound waves to create subsurface images, helping identify geological formations and potential hydrocarbon reservoirs. Remote sensing technologies, such as satellite imagery and aerial photography, offer valuable information about the Earth's surface, aiding in the identification of geological features and surface anomalies. Additionally, geochemical analysis involves studying the chemical composition of rocks, soil, and water to detect hydrocarbon indicators, using techniques like gas chromatography and mass spectrometry on samples collected during field surveys [47].

To integrate sustainability into exploration practices, advancements focusing on minimizing environmental impact play a vital role. Low-impact seismic methods offer an alternative to traditional surveys, aiming to reduce disruptions to ecosystems and wildlife habitats. These methods utilize existing vibrations and noise through "passive seismic" or "ambient noise" surveys, minimizing the need for active energy sources. Advanced remote sensing technologies integrated with Geographic Information Systems (GIS) [48] facilitate efficient mapping and analysis of potential exploration sites, enabling comprehensive assessments of environmental factors, land use, and ecological sensitivities when selecting areas for exploration. Moreover, conducting thorough Environmental Impact Assessments (EIA) [49] before and during exploration activities helps evaluate potential impacts on ecosystems, biodiversity, water resources, and local communities. This information informs decision-making and the implementation of mitigation measures to minimize adverse effects. Furthermore, the utilization of advanced data analytics, machine learning, and artificial intelligence techniques in exploration enhances the accuracy and efficiency of data interpretation. These technologies facilitate faster identification of potential reservoirs and the optimization of exploration strategies, reducing the need for extensive physical surveys and minimizing environmental disturbances [50].

3.5.2 Drilling Fluids

Drilling fluids, commonly known as mud, play a vital role in oil and gas exploration and drilling operations. They serve multiple functions, including cooling and lubricating the drill bit, maintaining wellbore stability, and facilitating the removal of drill cuttings [51]. When it comes to selecting the most suitable drilling fluid, the choice between water-based fluids (WBF) and non-aqueous based fluids (NABF) becomes paramount. Among NABFs, oil-based fluids (OBF) have traditionally been popular, but their limitations have prompted the industry to explore the advantages of water-based alternatives. WBF consist primarily of water, supplemented with various additives and chemicals to enhance their performance. They offer several distinct advantages over oil-based fluids, positioning them as the preferred choice in numerous drilling operations. WBFs exhibit a superior safety profile compared to their oil-based counterparts. Conventional OBFs, based on diesel or crude oil, have low flashpoints, making them potential fire hazards. In contrast, WBFs do not carry the same fire risk, thereby reducing the likelihood of accidents and enhancing overall operational safety. This advantage proves particularly crucial in high-temperature environments or when drilling in areas with stringent safety regulations. Only WBFs are permitted to be discharged into the sea, making them the preferred choice for offshore drilling activities. Conversely, synthetic-based fluids (SBF) and oil-based fluids face restrictions in environmentally sensitive areas due to concerns surrounding their potential toxicity. Additionally, the disposal costs associated with WBFs are generally lower, as they can be discharged into the sea in compliance with environmental regulations [52, 53].

In addition to the advantages of water-based fluids, the oil and gas industry is actively exploring various sustainable drilling techniques and approaches. Managed Pressure Drilling (MPD) is a drilling technique that enables precise control of wellbore pressure, reducing the risk of formation damage and enhancing drilling efficiency [54]. In MPD, the pressure in the wellbore is actively managed to maintain it within a narrow range that prevents undesirable conditions such as formation fluid influx or lost circulation. This is achieved by adjusting the surface backpressure, which can be done using specialized equipment such as rotating control devices (RCDs) and choke valves. By maintaining a controlled wellbore pressure, MPD minimizes the need for excessive mud circulation. This results in lower fluid consumption and reduced waste generation, leading to cost savings and a smaller environmental footprint [55]. Additionally, MPD enables drilling in challenging formations, such as those with narrow pressure margins or depleted reservoirs, where conventional drilling methods may face difficulties.

Underbalanced Drilling (UBD) is a technique where the wellbore pressure is intentionally kept lower than the formation pressure. This approach minimizes fluid losses and formation damage, leading to a reduced environmental impact and improved well productivity. By maintaining a lower wellbore pressure, UBD allows for controlled influx of reservoir fluids, which can enhance drilling efficiency and reduce formation damage caused by excessive mud invasion. UBD is particularly beneficial in formations with low-pressure reservoirs or those prone to fluid influx. It can help maximize well productivity and reduce formation damage by preventing the invasion of drilling fluids into the formation, which can impair reservoir connectivity and hinder hydrocarbon production. UBD also offers advantages in reducing formation damage associated with differential sticking and minimizing drilling-induced fractures [56].

Continuous circulation systems are designed to enable uninterrupted drilling operations, eliminating the need to halt mud circulation during critical operations such as making connections or tripping pipe. This approach minimizes the risk of fluid losses and enhances drilling efficiency while minimizing environmental impact. In a continuous circulation system, special equipment such as top drive or downhole circulation tools are used to maintain the flow of drilling fluids while making connections or performing other operations that traditionally required stopping mud circulation. By keeping the fluid flowing, this system reduces the risk of fluid losses into the formation, which can lead to well control issues and environmental concerns. It also improves drilling efficiency by minimizing downtime associated with stopping and restarting mud circulation [57]. Continuous circulation systems are particularly beneficial in challenging drilling environments, such as highly permeable formations or areas with narrow margins between pore pressure and fracture pressure. They help maintain wellbore stability, optimize drilling performance, and minimize the overall environmental impact of drilling operations [58].

3.5.3 Well Completion and Production

In the well completion stage, sustainable practices can be implemented by using eco-friendly materials for steel casing. For instance, utilizing recycled steel or steel with a low-carbon footprint can significantly reduce the environmental impact of casing manufacturing. Furthermore, sustainable cementing practices involve the use of environmentally friendly cement additives and optimizing the cementing process to minimize water and energy consumption. Examples include employing alternative cements such as geopolymers or low-carbon cements, which effectively reduce the carbon footprint associated with cement production. During the production stage, sustainable techniques focus on various aspects. Implementing sustainable pressure management techniques is vital to optimize production while minimizing environmental impact. An example is the injection of produced water or captured CO2 back into the reservoir, achieving both enhanced oil recovery and carbon sequestration. In terms of artificial lift techniques, the adoption of energy-efficient solutions is key [59]. For instance, utilizing electric submersible pumps (ESPs) powered by renewable energy sources like solar or wind helps reduce greenhouse gas emissions and dependency on fossil fuel-generated electricity [60].

Sustainable production practices also involve effective management and treatment of produced water. Technologies such as water recycling, membrane filtration, and advanced treatment processes play a crucial role in reducing freshwater consumption and ensuring proper disposal or reuse of produced water. Additionally, adopting gas management practices aims to minimize flaring and methane emissions during oil production. Capturing associated gas for power generation, re-injection into the reservoir for enhanced oil recovery, or conversion into value-added products like liquefied natural gas (LNG) are examples of sustainable approaches. Lastly, optimizing equipment efficiency is a key aspect of sustainable production. This can be achieved through the utilization of monitoring systems, predictive maintenance strategies, and energy-efficient equipment like high-efficiency motors and pumps [61].

Additionally, the implementation of rigorous casing integrity testing and monitoring protocols ensures long-term environmental protection by preventing fluid leakage into surrounding formations. In terms of sustainable production, the adoption of solar-powered ESPs in remote or off-grid locations is a practical example. These pumps harness renewable energy sources, reducing reliance on conventional power and lowering greenhouse gas emissions.

3.5.4 Surface Processing, Storage and Transportation

Chevron's implementation of energy efficiency measures resulted in a remarkable 35% reduction in greenhouse gas emissions. Similarly, water conservation techniques such as closed-loop systems and advanced water treatment technologies have enabled companies like BP to recycle and reuse process water, minimizing freshwater usage and demonstrating a commitment to responsible water management. In surface processing, the industry has successfully implemented emission control technologies like vapor recovery units (VRUs), flue gas desulfurization (FGD) systems, and selective catalytic reduction (SCR) systems. These measures effectively reduce volatile organic compounds (VOCs) and sulfur dioxide (SO_2) emissions, contributing to improved air quality in surrounding communities [62].

In storage and transportation, advancements in leak detection technologies such as fiber optic sensing systems and infrared-equipped drones have revolutionized the accuracy and efficiency of leak detection. Companies like Shell are utilizing fiber optic sensing systems to monitor pipelines and promptly address leaks, minimizing the potential for spills and environmental harm. Pipeline integrity management through regular inspections, preventive maintenance, and the use of advanced inspection tools like inline inspection tools ("smart pigs") ensures the safe and reliable transportation of oil. Companies are embracing green transport solutions, including the use of pipeline networks and electric/hybrid-powered vehicles, to reduce carbon emissions and enhance operational efficiency. Optimizing transport logistics, efficient loading and unloading practices, and minimizing travel distances further contribute to sustainability efforts.

Corrosion-resistant alloys like stainless steels, duplex stainless steels, and nickel-based alloys are being investigated for their exceptional resistance to corrosion in harsh environments, providing enhanced durability and longevity to infrastructure. Composite materials such as fiberglass reinforced plastic (FRP) and carbon fiber reinforced polymer (CFRP) offer high strength-to-weight ratios and outstanding corrosion resistance, making them ideal for construction in corrosive environments. Advanced coatings and linings, such as epoxy, polyurethane, and fusion-bonded epoxy (FBE), create a protective barrier between metal surfaces and corrosive elements, significantly reducing the risk of corrosion. Researchers are also exploring advanced cathodic protection systems, including impressed current cathodic protection (ICCP) and sacrificial anode systems, to supply protective currents and prevent corrosion. Additionally, corrosion-resistant design practices, such as optimized geometry and layout, incorporation of corrosion-resistant materials in critical areas, and proper coating and lining systems, play a vital role in minimizing the risk of corrosion initiation and propagation, ensuring extended service life for storage tanks and pipelines [63].

3.5.5 Other Practices

Certain additives and compounds used in oil refining pose risks to the environment and human health. Heavy metals like chromium, particularly in hexavalent form, are toxic and carcinogenic. Barite, a commonly used additive, can contain heavy metal impurities, leading to new regulations proposing the exclusion of contaminated sources. Additionally, certain aromatic hydrocarbons, fatty amine compounds, H2S scavengers, chelating agents, crosslinkers, and phosphorus-containing additives are associated with poor biodegradability, toxicity, or persistence, raising concerns about their environmental impact. The Harmonised Offshore Chemical Notification Format (HOCNF) [64] is a standardized format for exchanging information on chemicals in offshore oil and gas operations. It ensures consistent reporting and regulatory compliance. Similar datasheets and formats in the offshore industry include Safety Data Sheets (SDS), Chemical Safety Reports (CSR), European Chemicals Agency (ECHA) [65] Registration Dossiers, Material Safety Data Sheets (MSDS), and the Offshore Chemical Notification Scheme (OCNS) [66] used in the United Kingdom. These provide detailed information on properties, hazards, and safe handling of chemicals, ensuring safety and regulatory adherence in offshore operations.

In decommissioning, environmental impact assessments are crucial to identify risks and develop mitigation measures. Responsible waste management ensures the proper handling and disposal of decommissioning waste, preventing soil and water contamination. Habitat restoration and conservation efforts help mitigate ecological impact by rehabilitating disturbed areas and promoting biodiversity. Stakeholder engagement fosters transparency, inclusivity, and social responsibility throughout the decommissioning process, ensuring sustainable development and positive community outcomes.

Note

PLONOR: Positive List of Chemicals Not Requiring Further Assessment.

OSPAR: Convention for the Protection of the Marine Environment of the North-East Atlantic.

The PLONOR list, maintained by OSPAR, consists of chemicals that are deemed to pose little or no risk to the environment. The inclusion of a substance on the PLONOR list signifies that its use and discharge do not require stringent regulation. To be classified under the 'Green' category in the PLONOR list, a substance must meet specific criteria in terms of bioaccumulation, biodegradation, and toxicity. In terms of bioaccumulation, a substance on the PLONOR list should have a logarithm of the octanol–water partition coefficient (log POW) less than 3, a bioconcentration factor (BCF) less than 100, or a molecular weight greater than 700 g/mol. This indicates that the substance has low potential to accumulate in organisms. Biodegradability is another important criterion for PLONOR classification. Substances on the list should be readily biodegradable, meaning they can be broken down by natural processes in the environment. Additionally, the toxicity of a substance is considered. For a substance to be included in the PLONOR list, its lethal concentration (LC50) or effective concentration (EC50) should exceed 100 mg/l. This threshold ensures that the substance has low acute toxicity to aquatic organisms [67].

Examples of substances classified as PLONOR (posing little or no risk to the environment) include several inorganics, such as acetic acid, ethanol, butanol, propanol, glycerine, attapulgite clay, bentonite, cellulose, lignite, sodium lignosulfonate, calcium lignosulfonate, iron lignosulfonate, starch without additives, sugarcane molasses, and xanthan gum. These substances have been assessed and found to have minimal environmental impact based on their properties and behavior. By identifying substances that meet these criteria, the PLONOR list helps guide regulatory decisions and simplifies the management of chemicals in terms of environmental risk assessment and regulation, focusing resources on substances of greater concern while allowing for less stringent regulation of those posing little or no risk to the environment [68].

3.5.6 Strategies in Steel Manufacturing

In the raw material acquisition stage of steel manufacturing, sustainable practices involve utilizing scrap steel as a feedstock instead of relying solely on virgin raw materials. This reduces the demand for virgin resources, conserves natural resources and energy, and promotes responsible sourcing. In the ironmaking process, sustainable practices focus on optimizing blast furnace operations to reduce energy consumption and greenhouse gas emissions. The use of direct reduced iron (DRI) produced from recycled steel and improving blast furnace efficiency further enhance sustainability. For steelmaking, the adoption of electric arc furnace (EAF) technology, which utilizes recycled scrap steel, is more energy-efficient and emits fewer greenhouse gases compared to traditional methods. Maximizing the use of recycled scrap steel and increasing the share of steel produced through EAF technology promotes resource efficiency and reduces waste. Casting and rolling processes aim to minimize material loss and maximize yield through advanced process control technologies and continuous casting methods [26, 69].

Several companies, including ArcelorMittal, Nippon Steel Corporation, Tata Steel, and POSCO, are committed to reducing CO2 emissions in the steel industry through various measures such as low-carbon technologies, energy efficiency improvements, and hydrogen-based steelmaking. Additionally, global initiatives like ResponsibleSteel, World Steel Association's Climate Action Program, Climate

Group's SteelZero Initiative, and the Mission Possible Partnership are driving collaboration and promoting sustainable practices to achieve net-zero emissions and reduce the environmental impact of steel production. These efforts demonstrate a collective commitment to decarbonize the steel industry and foster sustainability across the supply chain [70].

3.6 Sustainable Reporting for Oil and Gas and Steel Industries

In the oil and gas industry, reporting practices include Environmental Performance Reports, Methane Emissions Reporting, Energy Efficiency Index, and Social Impact Assessments. These reports provide insights into environmental impacts, methane emissions, energy efficiency, and social contributions of oil and gas activities. For the steel industry, reporting practices involve Carbon Footprint Reports, Water Efficiency Metrics, Health and Safety Performance Reports, and Supply Chain Sustainability Assessments [71]. These reports track carbon emissions, water consumption, safety performance, and supply chain sustainability within the steel industry. Together, these reporting practices promote transparency, accountability, and sustainability in the oil and gas and steel sectors [72]. Figure 3.6 and Table 3.2 examines the standard parameters and impact of sustainability on the oil and gas industry.

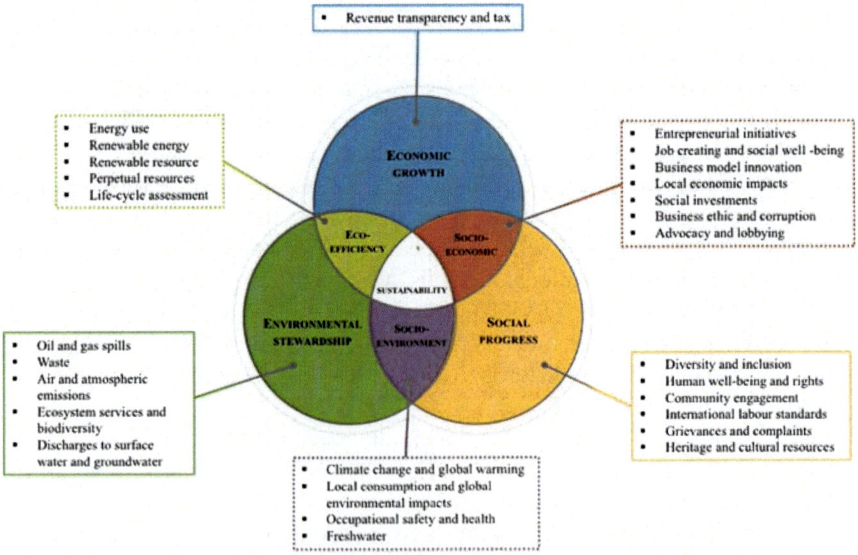

Fig. 3.6 Assessing the influence of sustainability on the oil and gas industry [73]

Table 3.2 Indicators and parameters for sustainability reporting in Oil and Gas companies

Report	Indicators and parameters	Metrics	Examples of companies
Environmental performance report	Greenhouse gas emissions	CO_2e emissions (metric tons)	Shell, ExxonMobil
	Methane emissions	Methane emissions (metric tons)	BP, Chevron
	Nitrous oxide emissions	Nitrous oxide emissions (metric tons)	Total, ConocoPhillips
	VOC emissions	VOC emissions (metric tons)	Equinor, Eni
	Water consumption	Water consumed (cubic meters)	Saudi Aramco, Gazprom
	Waste generation	Total waste generated (metric tons)	Repsol, Hess
	Air emissions	Pollutant-specific emissions (metric tons)	Occidental Petroleum, Woodside Petroleum
Methane emissions reporting	Methane leak rate	Number of methane leaks per unit of production	Chevron, Total
	Leak detection and repair programs	Number of leaks detected and repaired	ExxonMobil, Equinor
Energy efficiency index	Energy consumed	Energy consumed (megajoules) per barrel of oil equivalent (BOE)	Shell, BP
	Production of oil or gas	Barrels of oil equivalent (BOE) produced	Chevron, Total
Social impact assessment	Community engagement	Number of community engagement initiatives	Equinor, Repsol
	Local employment	Number of local employees hired	Saudi Aramco, Gazprom
	Social development projects	Contributions to social development projects (financial value)	Woodside Petroleum, Hess
Carbon footprint report	Scope 1 emissions (direct)	CO_2e emissions from direct sources (metric tons)	ArcelorMittal, Tata Steel
	Scope 2 emissions (indirect)	CO_2e emissions from purchased electricity (metric tons)	POSCO, Nippon Steel

(continued)

Table 3.2 (continued)

Report	Indicators and parameters	Metrics	Examples of companies
	Scope 3 emissions (indirect)	CO_2e emissions from value chain activities (metric tons)	China Baowu Steel, JFE Steel
Water efficiency metrics	Water consumption	Water consumed (cubic meters)	Nucor Corporation, Hyundai Steel
	Water recycling rates	Percentage of water recycled from total water consumption	Thyssenkrupp, JSW Steel
Health and safety performance report	Lost-time injury rate	Number of lost-time injuries per hours worked (per 100,000 h)	Tenaris, Suncor Energy
	Total recordable injury rate	Number of recordable injuries per hours worked (per 100,000 h)	ENOC Group, Marathon Petroleum
	Near-miss incidents	Number of near-miss incidents reported	Schlumberger, Baker Hughes
	Safety training programs	Number of safety training programs conducted	Halliburton, Weatherford International
Supply chain sustainability assessment	Responsible sourcing of raw materials	Percentage of raw materials from responsible and sustainable sources	Rio Tinto, BHP
	Labor and human rights standards	Compliance with labor and human rights standards	Vale, Freeport-McMoRan
	Supplier diversity	Percentage of suppliers from diverse backgrounds	Anglo American, MMK

3.7 Conclusion

In conclusion, the oil and gas industry and the steel industry are recognizing the growing importance of sustainability in their operations. Environmental impact and climate change concerns, resource depletion, and conservation, as well as social responsibility and stakeholder expectations, are driving the need for sustainable practices. Despite the presence of current initiatives, barriers to implementation still exist. However, strategies to promote sustainability are being developed and implemented across various stages of the oil and gas production process, including exploration, drilling fluids, well completion, production, and surface processing. Similarly, the steel industry is adopting strategies to minimize its environmental footprint. Sustainable reporting plays a crucial role in monitoring and communicating progress. Moving forward, it is imperative for these industries to continue prioritizing sustainability and collaborating to address challenges and drive positive change. By embracing sustainable practices, the oil and gas and steel industries can contribute to a more environmentally responsible and socially conscious future.

Activity

Multiple Choice Questions

Which of the following is a key driver for implementing sustainable strategies in the oil and gas and steel industries?

(a) Short-term financial gains
(b) Long-term environmental stewardship
(c) Reduction in production capacity
(d) Ignoring regulatory requirements

Which of the following is an example of a sustainable practice in the oil and gas industry?

(a) Excessive flaring of natural gas
(b) Neglecting environmental impact assessments
(c) Implementing water recycling systems
(d) Increasing reliance on coal for energy production

What is the primary goal of sustainable strategies in the steel industry?

(a) Maximizing carbon emissions
(b) Minimizing waste generation
(c) Ignoring employee safety
(d) Expanding production without limits

Which of the following is a benefit of sustainable strategies in the oil and gas industry?

(a) Increased air pollution
(b) Negative impact on local communities
(c) Reduced environmental footprint
(d) Ignoring climate change impacts

In the context of the steel industry, what does the term "circular economy" refer to?

(a) Maximizing resource depletion
(b) Promoting responsible waste management
(c) Minimizing worker benefits
(d) Ignoring product quality standards

Which of the following practices promotes sustainability in the oil and gas industry?

(a) Ignoring energy efficiency measures
(b) Expanding offshore drilling operations
(c) Implementing carbon capture and storage technologies
(d) Increasing reliance on non-renewable resources

What role can technology play in promoting sustainability in the steel industry?

(a) Increasing greenhouse gas emissions
(b) Minimizing waste generation
(c) Ignoring worker safety regulations
(d) Disregarding product quality standards

How can the oil and gas industry contribute to biodiversity conservation?

(a) Clearing natural habitats for exploration activities
(b) Implementing environmental monitoring programs
(c) Disregarding endangered species protections
(d) Expanding drilling operations in protected areas

Which of the following is a sustainable practice in the steel industry?

(a) Increasing water consumption
(b) Maximizing energy wastage
(c) Adopting cleaner production technologies
(d) Disregarding worker health and safety

What does the term "responsible sourcing" mean in the context of the oil and gas industry?

(a) Ignoring the environmental impact of raw material extraction
(b) Reducing employment opportunities in local communities
(c) Ensuring ethical and sustainable procurement practices
(d) Expanding operations in environmentally sensitive areas

How can the steel industry reduce its greenhouse gas emissions?

(a) Increasing the use of coal in production
(b) Implementing energy-efficient technologies
(c) Ignoring emission reduction targets
(d) Disregarding the use of renewable energy sources

Which of the following is an example of sustainable water management in the oil and gas industry?

(a) Overusing water in extraction processes
(b) Ignoring wastewater treatment
(c) Implementing water recycling systems
(d) Discharging untreated wastewater into rivers

What is the primary purpose of environmental impact assessments in the oil and gas industry?

(a) To expedite the extraction process
(b) To ignore potential environmental risks
(c) To minimize the impact on local ecosystems
(d) To disregard community concerns

How can the steel industry contribute to sustainable development?

(a) Ignoring social responsibility
(b) Minimizing energy efficiency
(c) Promoting responsible sourcing of raw materials
(d) Increasing reliance on fossil fuels

Which of the following is a potential benefit of sustainable practices in the oil and gas industry?

(a) Increased greenhouse gas emissions
(b) Negative impact on worker health and safety
(c) Enhanced product quality
(d) Reduced focus on recycling

What is the primary objective of sustainable strategies in the oil and gas industry?

(a) Ignoring environmental regulations
(b) Maximizing profits at any cost
(c) Minimizing negative environmental impacts
(d) Expanding carbon emissions

How can the steel industry reduce water consumption?

(a) Increasing water-intensive production processes
(b) Ignoring water conservation measures
(c) Implementing recycling and reusing techniques
(d) Discharging untreated wastewater into rivers

Which of the following is an example of sustainable waste management in the oil and gas industry?

(a) Dumping hazardous waste into open pits
(b) Ignoring waste reduction measures
(c) Implementing proper waste disposal and recycling practices
(d) Burning waste materials without any treatment

How can the oil and gas industry contribute to the United Nations Sustainable Development Goals (SDGs)?

(a) Disregarding social responsibility
(b) Minimizing collaboration with local communities
(c) Embracing renewable energy sources
(d) Expanding operations in environmentally sensitive areas

Which of the following practices promotes sustainability in the oil and gas industry?

(a) Ignoring community engagement
(b) Increasing reliance on non-renewable resources
(c) Implementing energy efficiency measures
(d) Disregarding climate change impacts

Answers:

b) Long-term environmental stewardship.
(c) Implementing water recycling systems.
(b) Minimizing waste generation.
(c) Reduced environmental footprint.
(b) Promoting responsible waste management.
(c) Implementing carbon capture and storage technologies.
(b) Minimizing waste generation.
(b) Implementing environmental monitoring programs.
(c) Adopting cleaner production technologies.
(c) Ensuring ethical and sustainable procurement practices.
(b) Implementing energy-efficient technologies.
(c) Implementing water recycling systems.
(c) To minimize the impact on local ecosystems.
(c) Promoting responsible sourcing of raw materials.
(c) Enhanced product quality.
(c) Minimizing negative environmental impacts.
(c) Implementing recycling and reusing techniques.
(c) Implementing proper waste disposal and recycling practices.
(c) Embracing renewable energy sources.
(c) Implementing energy efficiency measures.

References

1. L. M. M. Araújo, C. B. S. Maior, I. D. Lins and M. J. das C. Moura, *Geoenergy Science and Engineering*, 2023, **226**, 211771.
2. R. Kumar and T. Markeset, *Journal of Business & Industrial Marketing*, 2007, **22**, 272–280.
3. How the Oil and Gas Industry Works, https://www.investopedia.com/investing/oil-gas-industry-overview/, (accessed July 21, 2023).
4. S. Saraji and D. Akindipe, in *Sustainable Oil and Gas Using Blockchain*, eds. S. Saraji and S. Chen, Springer International Publishing, Cham, 2023, pp. 3–33.
5. Connotations of scientific development of giant gas fields in China - ScienceDirect, https://www.sciencedirect.com/science/article/pii/S2352854020300929?via%3Dihub, (accessed July 23, 2023).
6. N. Burclaff, Research Guides, https://guides.loc.gov/oil-and-gas-industry/introduction, (accessed July 21, 2023).
7. The Oil and Gas Industry in Energy Transitions – Analysis, https://www.iea.org/reports/the-oil-and-gas-industry-in-energy-transitions, (accessed July 21, 2023).

8. M. Cartwright, The Steel Industry in the British Industrial Revolution, https://www.worldhistory.org/article/2206/the-steel-industry-in-the-british-industrial-revol/, (accessed July 21, 2023).
9. N. Conte, Visualizing 50 Years of Global Steel Production, https://www.visualcapitalist.com/visualizing-50-years-of-global-steel-production/, (accessed July 23, 2023).
10. 1E. Hasani, P. Gelard and N. Azad, *Public Policy In Administration*, 2023, **13**, 23–50.
11. A. D'Costa, *The Global Restructuring of the Steel Industry: Innovations, Institutions and Industrial Change*, Routledge, 1999.
12. A. N. Conejo, J.-P. Birat and A. Dutta, *Journal of Environmental Management*, 2020, **259**, 109782.
13. W. Sun, Q. Wang, Y. Zhou and J. Wu, *Applied Energy*, 2020, **268**, 114946.
14. H. Dong, Y. Liu, L. Wang, X. Li, Z. Tian, Y. Huang and C. McDonald, *Ironmaking & Steelmaking*, 2019, **46**, 922–927.
15. M. Pei, M. Petäjäniemi, A. Regnell and O. Wijk, *Metals*, 2020, **10**, 972.
16. M. D. Anis and T. Z. Siddiqui, *J. Mgmt. & Sustainability*, 2015, **5**, 115.
17. A. Bhaskar, R. Abhishek, M. Assadi and H. N. Somehesaraei, *Journal of Cleaner Production*, 2022, **350**, 131339.
18. M. Sharma, S. Joshi, M. Prasad and S. Bartwal, *Journal of Cleaner Production*, 2023, **391**, 136133.
19. Z. Wang, S. Li, Z. Jin, Z. Li, Q. Liu and K. Zhang, *Energy Strategy Reviews*, 2023, **45**, 101048.
20. L. E. K. Holappa, in *Applications of Process Engineering Principles in Materials Processing, Energy and Environmental Technologies*, eds. S. Wang, M. L. Free, S. Alam, M. Zhang and P. R. Taylor, Springer International Publishing, Cham, 2017, pp. 401–410.
21. https://academic-accelerator.com/encyclopedia/carbon-emission-trading
22. STEEL – Carbon Smart Materials Palette, https://materialspalette.org/steel/, (accessed July 23, 2023).
23. M. M. (Marco M. Tivelli, Thesis, Massachusetts Institute of Technology, 2004.
24. A. S. Pascaris and J. M. Pearce, *Energies*, 2020, **13**, 3932.
25. M. Kano and Y. Nakagawa, *Computers & Chemical Engineering*, 2008, **32**, 12–24.
26. G. W. Watson, D. Scott, J. Bishop and T. Turnbeaugh, *J Bus Psychol*, 2005, **19**, 303–318.
27. How Aramco's circular economy thinking can help the planet, https://www.aramco.com/en/magazine/elements/2021/circular-economy, (accessed July 21, 2023).
28. Reliance Annual Report 2019 - 2020, https://www.ril.com/ar2019-20/mda.html, (accessed July 21, 2023).
29. Shell Sustainability Report 2021 - Home, https://reports.shell.com/sustainability-report/2021, (accessed July 21, 2023).
30. Annual Reports 2020–21 - en - ongcindia.com, https://ongcindia.com/web/eng/investors/annual-reports/annual-reports21, (accessed July 21, 2023).
31. D. Naghi Beiranvand, K. Jamali Firouzabadi and S. Dorniani, *Journal of Modelling in Management*, 2021, **18**, 117–146.
32. M. Arena, G. Azzone, S. Ratti, V. M. Urbano and G. Vecchio, *Sustainable Development*, 2023, **31**, 12–25.
33. S. Pati, *Safety Science*, 2023, **159**, 106006.
34. Regulation and impact of VOC and CO2 emissions on low-carbon energy systems resilient to climate change: A case study on an environmental issue in the oil and gas industry - Fetisov - 2023 - Energy Science & Engineering - Wiley Online Library, https://onlinelibrary.wiley.com/doi/full/https://doi.org/10.1002/ese3.1383, (accessed July 21, 2023).
35. CCUS in Clean Energy Transitions – Analysis, https://www.iea.org/reports/ccus-in-clean-energy-transitions, (accessed July 21, 2023).
36. Y. Mahmood, T. Afrin, Y. Huang and N. Yodo, *Sustainability*, 2023, **15**, 4953.
37. Ramasubramanian, B., Reddy, M. V., Zaghib, K., Armand, M., & Ramakrishna, S. (2021). Growth mechanism of micro/nano metal dendrites and cumulative strategies for countering its impacts in metal ion batteries: A review. Nanomaterials, 11(10), 2476.

38. Analysis of iron and steel production paths on the energy demand and carbon emission in China's iron and steel industry | SpringerLink, https://link.springer.com/article/https://doi.org/10.1007/s10668-022-02234-5, (accessed July 21, 2023).
39. S. E. Barykin, S. M. Sergeev, A. B. Mottaeva, E. De La Poza Plaza, N. V. Baydukova and A. V. Gubenko, *Environ Sci Pollut Res*, 2023, **30**, 6849–6863.
40. Q. Yue, X. Chai, Y. Zhang, Q. Wang, H. Wang, F. Zhao, W. Ji and Y. Lu, *Environ Dev Sustain*, 2023, **25**, 4065–4085.
41. Ipieca | Advancing environmental and social performance across the energy transition, https://www.ipieca.org/, (accessed July 21, 2023).
42. M. Liu, Y. Li, Y. Xu, L. Chen, Q. Wang, Q. Ma and X. Yuan, *Journal of Cleaner Production*, 2023, **389**, 136048.
43. A multivariate quantitative approach for sustainability performance assessment: An upstream oil and gas company | SpringerLink, https://link.springer.com/article/https://doi.org/10.1007/s10668-022-02112-0, (accessed July 21, 2023).
44. Towards sustainability in the global oil and gas industry: Identifying where the emphasis lies - ScienceDirect, https://www.sciencedirect.com/science/article/pii/S2665972721000465, (accessed July 21, 2023).
45. S. Joshi, P. Jadhawar and A. Kumar, *Challenges and Recent Advances in Sustainable Oil and Gas Recovery and Transportation*, Elsevier, 2023.
46. J. Y. Abor, A. Karimu and R. Brännlund, *The Economics of the Oil and Gas Industry: Emerging Markets and Developing Economies*, Taylor & Francis, 2023.
47. T. Li, *Open Access Library Journal*, 2023, **10**, 1–10.
48. GIS (Geographic Information System), https://education.nationalgeographic.org/resource/geographic-information-system-gis, (accessed July 21, 2023).
49. Environmental impact assessment, https://environment.ec.europa.eu/law-and-governance/environmental-assessments/environmental-impact-assessment_en, (accessed July 21, 2023).
50. L. DOU, Z. WANG, F. GAO and X. LIU, *Bulletin of Chinese Academy of Sciences (Chinese Version)*, 2023, **38**, 59–71.
51. Drilling mud | Benefits, Types & Uses | Britannica, https://www.britannica.com/technology/drilling-mud, (accessed July 21, 2023).
52. Y. J. Amuda, S. Hassan and U. Subramaniam, *Energies*, 2023, **16**, 3151.
53. A. Rana, M. Murtaza, T. A. Saleh, M. S. Kamal and M. Mahmoud, *Journal of Petroleum Science and Engineering*, 2022, **214**, 110332.
54. C. (周长铖) Zhou, M. (何淼) He, M. (许明标) Xu, D. (林德举) Lin and X. (陈鑫) Chen, *Physics of Fluids*, 2023, **35**, 014115.
55. Z. Zhang, B. Sun, Z. Wang, X. Mu and D. Sun, *Energy*, 2023, **262**, 125619.
56. Review of underbalanced drilling techniques highlighting the advancement of foamed drilling fluids | SpringerLink, https://link.springer.com/article/10.1007/s13202-022-01596-w, (accessed July 21, 2023).
57. J. Ramasamy and M. K. Al-Arfaj, OnePetro, 2023.
58. Drilling Mud, https://www.investopedia.com/terms/d/drillingmud.asp, (accessed July 21, 2023).
59. M. Zeidouni, in *Shale Hydrocarbon Recovery: Basic Concepts and Reserve Estimation*, ed. M. Zeidouni, Springer International Publishing, Cham, 2023, pp. 11–27.
60. A. Petrochenkov, P. Ilyushin, S. Mishurinskikh and A. Kozlov, *Inventions*, 2023, **8**, 29.
61. F. Kong, Y. Liu, L. Tong, W. Guo, Y. Jin, L. Wang and Y. Ding, *Applied Thermal Engineering*, 2023, **233**, 121172.
62. T. Kutz, M. Davis, R. Creek, N. Kenaston, C. Stenstrom and M. Connor, *Interfaces*, 2014, **44**, 39–54.
63. E. W. Harnisch Lenny Bernstein, Joyashree Roy, Lynn Price, Jochen, in *Renewable Energy*, Routledge, 2011.
64. HOCNF | FUCHS LUBRICANTS NORWAY AS, https://www.fuchs.com/no/en/products/product-program/offshore-and-marine/hocnf/, (accessed July 21, 2023).

65. European Chemicals Agency – ECHA | European Union, https://european-union.europa.eu/institutions-law-budget/institutions-and-bodies/search-all-eu-institutions-and-bodies/european-chemicals-agency-echa_en, (accessed July 21, 2023).
66. Offshore Chemical Notification Scheme (OCNS) - Cefas (Centre for Environment, Fisheries and Aquaculture Science), https://www.cefas.co.uk/data-and-publications/ocns/, (accessed July 21, 2023).
67. What is the PLONOR-list?, https://www.borregaard.com/markets/oil-gas/sustainability/what-is-the-plonor-list/, (accessed July 21, 2023).
68. PLONOR Definition, https://www.lawinsider.com/dictionary/plonor, (accessed July 21, 2023).
69. I. Tolosa, F. Garciandía, F. Zubiri, F. Zapirain and A. Esnaola, *Int J Adv Manuf Technol*, 2010, **51**, 639–647.
70. P. K. Hajoary, *Procedia Computer Science*, 2023, **217**, 614–619.
71. F. Sotti and S. Santucci, *Risk Governance and Control Financial Markets & Institutions*, 2023, **13**, 52–66.
72. B. Gajdzik, R. Wolniak and W. Grebski, *Energies*, 2023, **16**, 3384.
73. N. Capobianco, V. Basile, F. Loia and R. Vona, *Sustainability*, 2021, **13**, 6266.

Chapter 4
Effective Waste Management Strategies and Circularity of Plastics

Abstract This chapter provides a examination of the circularity principles of plastics and their corresponding environmental impacts. It begins by exploring the importance of moisture control in plastic products and the negative effects that can arise from exposure to moisture. Additionally, it investigates the role of ash and carbon content in plastics and how they can be managed to minimize their impact on the environment. The chapter then proceeds to investigate the end-of-life options for plastics, such as landfill, incineration, and composting. Each option is evaluated in terms of its methodology and challenges. The chapter goes on to explore various waste recycling and upcycling technologies and their benefits. Additionally, the chapter covers emerging technologies such as microwave-assisted and plasma-assisted conversion, which offer new and innovative ways to recycle and upcycle plastics. The social and governance aspects of plastic waste are also examined, including the concept of extended producer responsibility (EPR) and its potential to motivate manufacturers to design more sustainable products that are easier to recycle. Relevant policies and schemes are also discussed, including the role of government regulation in promoting sustainable waste management practices.

Keywords Moisture control · Plasma-assisted · Upcycling · Life cycle assessment

4.1 Paradox of Plastic: Value Versus Lifespan

Plastic waste accumulation is one of the major environmental issues, with over 300 million tons of waste produced globally every year and their production has increased exponentially from 1950 to 2022 as shown in Fig. 4.1. Only 9% of plastic waste is recycled, and the rest ends up in landfills or the environment [1]. This has resulted in plastic pollution affecting marine wildlife, with an estimated 100,000 marine mammals and turtles, and 1 million seabirds dying every year due to plastic ingestion or entanglement [2]. In fact, there are currently over 5 trillion pieces of plastic in the oceans, with an estimated 8 million tons added every year. Even more concerning, microplastics (μm scale) that is found in the drinking water, sea salt, and even in the

S. Ramakrishna and B. Ramasubramanian, *Handbook of Materials Circular Economy*,
https://doi.org/10.1007/978-981-97-0589-4_4

atmospheric air. It is important to note that plastic waste costs around $13 billion a year in environmental damage to marine ecosystems alone [3]. Therefore, it is critical to effectively manage plastic waste and transition to a circular economy for plastics. To facilitate this, plastic waste can be broadly categorized into types such as single-use plastics, packaging plastics, construction plastics, electronic plastics, and automotive plastics. Additionally, consumer plastic waste are majorly categorized based on its composition, such as Polyethylene Terephthalate (PET), High-Density Polyethylene (HDPE), Polyvinyl Chloride (PVC), Low-Density Polyethylene (LDPE), and Polypropylene (PP) Fig. 4.2a. Understanding these categories is key to developing effective waste management strategies that help mitigate the damage caused by plastic waste. The global plastic pollution with higher landfill, as shown in Fig. 4.2b is mostly in Asian countries.

The idea of plastic value is complicated, and its longevity and effects must be carefully considered. Understanding that not all plastics are created equal and that various kinds have differing lifespans and environmental impacts is crucial. For instance, it might take up to 450 years for PET to decompose, a common plastic used in bottles and food packaging [4]. Like HDPE, which is used in milk jugs and detergent bottles, it may take up to 1000 years to decompose. However, under certain circumstances, some plastics, including polylactic acid (PLA), manufactured from renewable materials like corn-starch, disintegrate in minimum period of six months. However, PLA is not readily recyclable and, if not disposed of, might contaminate other polymers [5, 6].

The zero-waste idea acknowledges that trash is an asset in transformation that is produced during the transitional periods between manufacturing and consumption.

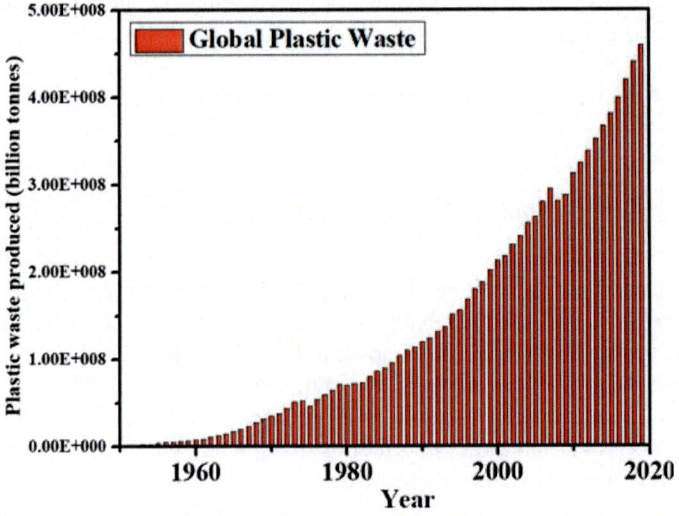

Fig. 4.1 Global plastic pollution from 1950 to 2020 with exponential increase to 4.48×10^8 billion tonnes. Data sourced from Our world in data

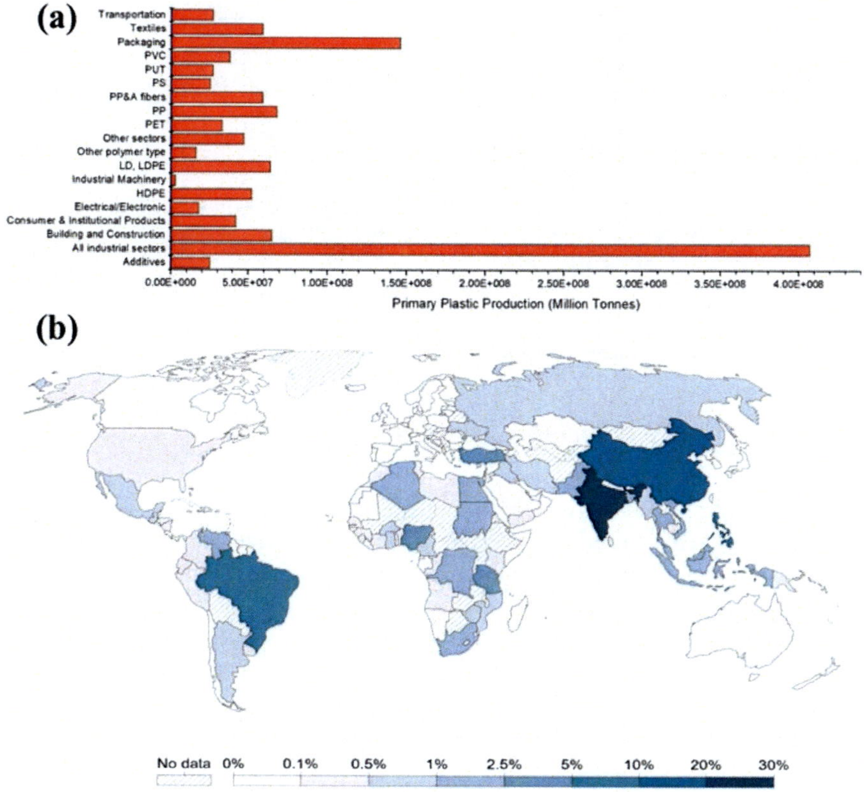

Fig. 4.2 **a** Primary plastic production in million tonnes with identified major contributors as textiles, building and construction, packaging (HDPE, PP). **b** Global plastic consumption and low recycling rates. Data sourced from Statistica and Our world in data

The preservation and maintenance of resource value is the basic goal of zero-waste. The value of items and their assets is demolished as a consequence of conventional waste control systems' prioritization of safe garbage disposal. Rather, zero-waste prioritizes the top five waste reduction strategies, including reduction, reuse, restoring, recycling, and reselling. The efficient and secure reuse of plastic is one challenge that emerges in the context of zero-waste. As plastic's micro-particle breakdown products are extremely hazardous, it is uncertain whether plastic can be managed under the zero-waste goal or if it falls under the hazardous waste goal. Value is added throughout a product's life cycle, from resource extraction and refining through production and distribution. The value of post-use items is reduced by traditional waste management methods, which recover materials at the lowest level. In contrast, the zero-waste strategy emphasizes on the circular economy model to preserve as high value after usage, Fig. 4.3. The goal of this concept is to minimize waste and keep products and supplies in circulation for as long as feasible [7, 8] (Table 4.1).

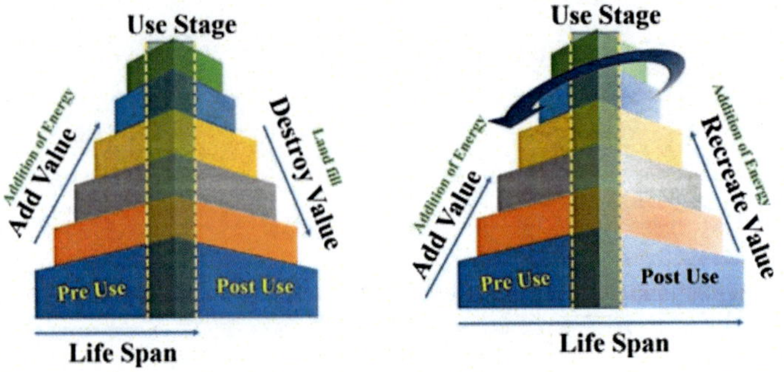

Fig. 4.3 Value of plastic with respect to lifespan (recreate plastics post usage to increase the value)

4.1.1 Circularity Principles of Plastics

Circularity in plastics is a concept that aims to keep plastic materials in use for as long as possible through a closed-loop system of recycling, reuse, and waste reduction. The goal of circularity is to minimize waste and maximize resource efficiency, in alignment with the principles of the SDGs. The circular economy seeks to design products and processes that keep materials and resources in use for as long as possible, while reducing waste and minimizing the environmental impact of production and disposal.

Four principles are centralized to achieving circularity in plastics. The first principle is to reduce the amount of plastic waste generated by minimizing the use of single-use plastics and designing products that use less plastic overall. The second principle is to promote the reuse of plastic products and packaging wherever possible, such as through refillable bottles or reusable shopping bags. The third principle involves promoting the recycling of plastic waste by designing products that are easily recyclable, improving recycling infrastructure, and increasing consumer awareness of recycling options. The fourth principle is to recover the energy and resources from plastic waste through methods such as waste-to-energy or chemical recycling (Fig. 4.4).

4.1.2 Moisture Control in Plastic

The amount of moisture present, significantly affect the recycling of plastic. Plastic waste has moisture content of less than 1%, yet this relatively modest quantity of moisture is troublesome for plastic manufactures since it results in considerable polymer breakdown. Moisture content in food packaging polymers should be less than 0.5% to prevent bacterial growth and maintain food safety. Medical-grade polymers must be free of moisture to prevent contamination and maintain product purity.

Table 4.1 Structure, applications and commonly employed recycling methods for plastic polymer

Plastic polymer	Structure	Applications	Recycling method	Recycling ease
Polyethylene (PE)	$-(-CH_2-CH_2)-$	Plastic bags, packaging, bottles, toys, pipes	Mechanical recycling, pyrolysis, gasification	Easy
Polypropylene (PP)	$-(-CH_2\text{-}CH(CH_3)-)-$	Packaging, automotive parts, toys, medical devices	Mechanical recycling, pyrolysis, gasification	Easy
Polyvinyl chloride (PVC)	$-(-CH_2-CHCl-)-$	Pipes, flooring, window frames, medical tubing, electrical cable insulation	Chemical recycling, mechanical recycling, pyrolysis, gasification	Difficult
Polystyrene (PS)	$-(-CH_2\text{-}CH(C_6H_5)-)-$	Foam products, packaging, disposable cups and plates	Mechanical recycling, pyrolysis, gasification	Difficult
Polyethylene terephthalate (PET)	$-(-OC_6H_4COO-)n-$	Bottles, food containers, clothing, carpet fibers	Mechanical recycling, chemical recycling	Easy
Polyurethane (PU)	$-(-NHCOO-)_n-$	Foam products, adhesives, coatings, insulation	Mechanical recycling, chemical recycling, pyrolysis, gasification	Difficult
Polycarbonate (PC)	$-(-O-C_6H_4-CO-C_6H_4-O-)_n\text{-}$	CDs, lenses, medical devices, water bottles	Mechanical recycling, pyrolysis, gasification	Difficult
Acrylonitrile–butadiene–styrene (ABS)	$-(C_8H_8)_x-(C_4H_6)_y-(C-H-N)_z-$	Lego bricks, automotive parts, appliances, toys	Mechanical recycling	Easy
Polyethylene-vinyl acetate (PEVA)	$-(-CH_2-CH_2-OCOCH_3-)-$	Shower curtains, raincoats, inflatable toys	Mechanical recycling, incineration	Easy
Polybutylene terephthalate (PBT)	$-(-OC_6H_4COO-(CH_2)_4O-)_n-$	Electrical connectors, automotive parts, small appliances	Mechanical recycling	Easy
Polymethyl methacrylate (PMMA)	$-(-O-CO-C(CH_3)=CH_2-)_n\text{-}$	Windows, signs, lighting, aquariums	Mechanical recycling, pyrolysis, gasification	Difficult

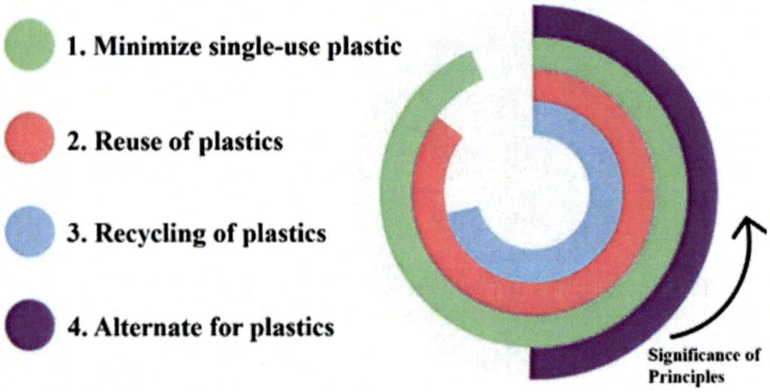

Fig. 4.4 Four principles for circularity in plastics

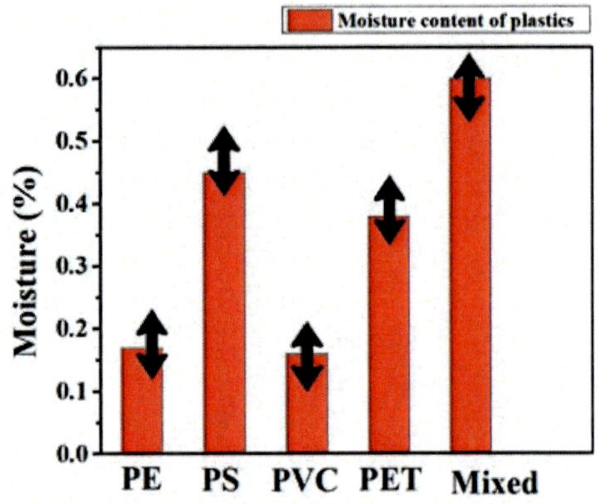

Fig. 4.5 Moisture content in the commonly employed plastic polymers. Data sourced from Statistica and Our world in data

High moisture levels in polymers used in electronics and electrical applications leads to corrosion, and electrical shorts [9, 10] (Fig. 4.5).

Moisture may have several detrimental consequences on the recycling of plastic, including lowering the value of the recycled plastic, generating processing issues, and perhaps causing machinery damage. Moisture content degrades the recycled plastic's quality. Moisture may result in flaws including bubbles, voids, and surface irregularities, which reduce the plastic's mechanical strength. In addition, steam generated during processing corrodes the recycling equipment and cause malfunctioning.

4.1.3 Ash and Carbon Content

Plastics typically have low ash content, which makes them challenging to recycle through incineration. Incineration is a process of burning waste materials to produce energy, and materials with high ash content are suitable for energy recovery through this process. However, the low ash content of most plastics makes them difficult to recycle through incineration, as the resulting ash is not valuable and requires disposal. In contrast, materials with a higher ash content, such as PS, can be suitable for energy recovery through incineration, where the heat generated by burning the material can be used for electricity or heating. Excessive amount of ash must be removed since it negatively impacts the pyrolysis performance in terms of bio-oil output. Plastic trash has a fixed carbon content ranging from 0 to 32%, with biochar being the carbon-rich porous matter that persists after devolatilization [11, 12]

The carbon content of plastics is an additional factor that significantly influences their recyclability. Plastics are primarily composed of hydrocarbons, which are carbon and hydrogen-based compounds. As a result, plastics have a high carbon content that can negatively impact their ability to be recycled. When subjected to mechanical recycling processes, plastics are melted and reprocessed to create new products. However, plastics with a high carbon content are susceptible to thermal degradation during the melting process, which adversely affect the properties and lead to a reduction in the strength or increased brittleness of the recycled plastic compared to the original material [12].

4.2 End-of-Life Plastics

Consumer plastics are typically made of six distinct kinds of polymer resins. Each resin type is identified by a resin identification number, which ranges from 1 to 7, that is moulded or stamped onto the surface of the plastic goods (Fig. 4.6).

HDPE and LDPE are commonplace in the packaging industry due to their exceptional durability and versatility. HDPE and LDPE are highly durable and withstand exposure to chemicals, moisture, and other environmental factors. LDPE is flexible and malleable than HDPE, making it ideal for products that require flexibility, such as squeeze bottles. Additionally, both HDPE and LDPE are lightweight, making them

Fig. 4.6 Common plastics and their symbol terminologies. Reproduced with copyrights from [13], 2020, CC BY

suitable for packaging products that need to be transported over long distances. HDPE accounts for approximately 16% of all plastic packaging produced globally, while LDPE accounts for about 12% [14]. When combined, HDPE and LDPE constitute almost 30% of all plastic packaging generated worldwide.

4.2.1 Landfill

The majority of the plastic trash in the landfills of the European Union in 2018 was made up of packaging plastics such HDPE, LDPE, PP, PET, PS, and PVC. The projected annual usage of these plastics was 90.0 ∓ 4.8 kg per capita in 2016, with PVC accounting for 16 kg and PET for 68.0 ∓ 4.8 kg [15]. In order for landfills to function, a variety of biological, chemical, and physical processes must take place in order to transform solid waste and water into gas and leachate. These procedures may be divided into five phases: early adaptation, transition, creation of acids, methanol fermentation, and ultimate maturation and stability of solid waste. Plastics degrade in a variety of ways during these phases, involving aerobic decay, acid production. However, there are still questions about the long-term destiny of plastics in landfills due to worries about their recalcitrance, biodegradation, or disintegration.

To mitigate the impact of landfills, it is important to segregate waste materials into different categories. This process is known as waste segregation. Different types of waste require different treatment methods. For example, plastic waste can be recycled, while organic waste can be composted. Segregating waste also helps to reduce the amount of waste that ends up in landfills, reducing the environmental impact. In many countries, different coloured bins are used to indicate the type of waste that should be deposited in them. For instance, in the United Kingdom, the green bin is used for garden waste, brown bin is used for food waste, blue bin is used for paper, and black bin is used for non-recyclable waste. In some areas, a red bin is used for hazardous waste such as batteries and chemicals. The use of coloured bins helps to make waste segregation easier for people, promoting proper disposal practices and reducing the amount of waste that ends up in landfills [16] (Table 4.2).

The formula used to calculate the methane release from plastics is:

$$\begin{aligned}\text{Methane release (kg)} = {} & \text{Amount of plastic in landfill (kg)} \\ & \times \text{Methane generation potential (MGP) of plastic} \left(\text{kg CH}_4/\text{kg plastic}\right) \\ & \times \text{Methane conversion factor (MCF) of landfill}\end{aligned}$$

The MGP of plastic refers to the amount of methane gas produced per unit weight of plastic in a landfill. The MCF of landfill refers to the efficiency of the landfill in capturing and controlling methane emissions.

Table 4.2 List of common plastics polymers with threat level, methane gas emission in GWP and alternates

Plastic name	Landfill level of threat	Estimated methane gas emission in GWP (kg CO_2eq/kg)	Alternate material with low threat
PET (polyethylene terephthalate)	Low	22–28	Glass, aluminum
HDPE (high-density polyethylene)	Low	34–44	Paper, biodegradable plastic
PVC (polyvinyl chloride)	High	80–100	Glass, wood, metal
LDPE (low-density polyethylene)	Low	42–54	Paper, biodegradable plastic
PP (polypropylene)	Low	56–72	Glass, bamboo
PS (polystyrene)	Moderate	73–92	Paper, biodegradable plastic
Other plastics	Varies	Varies	Biodegradable plastic, compostable material, glass, metal

Note The estimated methane gas emissions in GWP are based on the decomposition of plastics in landfills, and the values are provided in terms of the equivalent amount of carbon dioxide (CO_2) emissions over a 100-year time horizon. The actual amount of methane produced may vary depending on landfill conditions

4.2.2 *Incineration*

Incineration is a process of burning plastic waste at very high temperatures, usually ranging between 850 and 1200 °C. This method releases energy in the form of heat, light, and gases. However, certain types of plastic, like PVC with high chlorine content, pose a serious incineration threat because they emit toxic gases and dioxins during combustion. For instance, PVC releases hydrochloric acid and other dangerous gases, while incomplete combustion of these plastics can produce dioxins, a highly toxic chemical. Similarly, PS also emit harmful gases during incineration [17]. The incineration process involves thermal decomposition, oxidation, and gasification mechanisms. In thermal decomposition, the plastic is heated to its melting point, and the polymer chains break down into smaller molecules. In oxidation, the smaller molecules are burned in the presence of oxygen, releasing heat and light energy. In gasification, the remaining solid residues are converted into a gas that can be used to generate energy [18].

The threats associated with incineration of plastics include the release of toxic gases, pollutants such as dioxins, furans, and heavy metals, which can have detrimental effects on human health and the environment. According to the U.S. Environmental Protection Agency (EPA), incineration is the second-largest source of dioxin emissions in the United States, accounting for 13% of total emissions in 2018 [19]. Moreover, incineration produces significant amounts of ash and other residues

Fig. 4.7 Waste plastic % incinerated, discarded and recycled from 1980 to 2015. Data sourced from Statistica and Our world in data

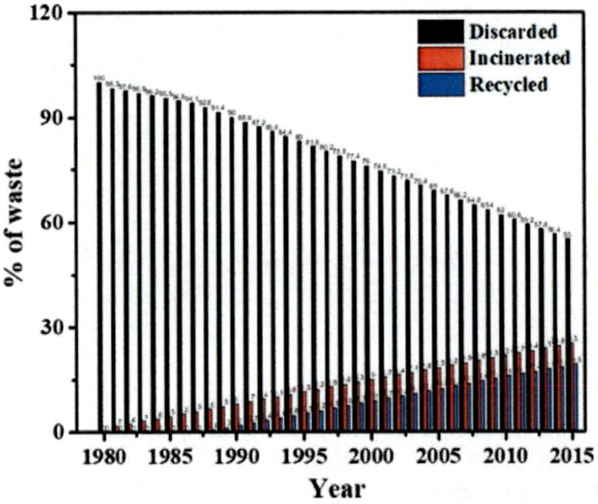

that may contain hazardous chemicals and require special disposal methods. The EPA reports that incineration generates more than 7 million tons of ash and other residues annually in the United States, and proper disposal is crucial to prevent toxic substances from contaminating the environment [19, 20] (Fig. 4.7).

Incinerated ashes are influenced by multiple factors, including the incineration process, the type of waste incinerated, and the incinerator used, which ultimately determine their physical, chemical, and mineralogical properties. The physical characteristics of incinerated ashes include particle size distribution, bulk density, and specific surface area. Chemical properties of incinerated ashes vary depending on the composition of the waste incinerated, but typically contain minerals, trace elements, and heavy metals. Mineralogical properties of incinerated ashes can be determined by XRD analysis [20]. Two main types of incinerated ashes are generated: bottom ash and fly ash. Bottom ash is the residue that settles at the bottom of the incinerator and is composed of larger particles with lower heavy metal concentrations than fly ash. Fly ash is the fine, powdery residue carried by hot gases and removed from the flue gas by filters or electrostatic precipitators. Fly ash has pozzolanic properties and is commonly used in the production of cement, concrete, and other construction materials. On the other hand, bottom ash can be processed for use as a construction material, such as in road building or as a substitute for aggregates in concrete [19].

Post-processing methods are employed to enhance the physical and chemical properties of incinerated ashes. Ash can be washed to eliminate residual contaminants or impurities, or it can be treated with stabilizing agents to prevent leaching of heavy metals. Additional materials, such as binders or aggregates, are added to improve the strength or workability of ash-based materials.

4.2.3 Composting

Composting is a process to break down biodegradable plastics, made from organic materials like corn-starch or sugarcane. However, the process requires careful management and specific conditions. Not all types of plastic are compostable, only biodegradable plastics, which can be broken down by microorganisms. The composting process involves creating an environment that supports the growth of microorganisms that break down organic matter. Moisture, oxygen, and temperature are essential factors for this process. Moisture keeps the compost pile from drying out, oxygen supports aerobic bacteria that break down the organic material, and temperature speeds up the decomposition process. The compost pile requires regular turning to mix the materials and provide oxygen to the microorganisms. The ideal temperature for composting is between 135 and 160 °F (57–71 °C), which helps speed up the decomposition process [21]. Additional dry materials like leaves or shredded newspaper can be added to cool the compost pile if it becomes too hot. It's important to note that composting plastic waste is not a quick process and can take several months or even years to decompose fully. Examples of biodegradable plastics include polylactic acid (PLA), polyhydroxyalkanoates (PHA), and starch-based plastics [21].

The carbon-to-nitrogen ratio is crucial for composting success. Carbon-rich materials (e.g., dry leaves) provide energy, while nitrogen-rich materials (e.g., food scraps) provide protein. A 25 to 30:1 carbon-to-nitrogen balance is optimal. Adequate moisture is essential (50–60%), but too much lowers incineration efficiency. Oxygen is necessary for aerobic bacteria to grow; insufficient oxygen leads to anaerobic conditions, odour, and reduced decomposition. Material size affects the decomposition rate; small pieces decompose faster but may restrict oxygen flow if too small (μm) [21].

The scope of the conceptual framework, Fig. 4.8, take into account not only the individual types of plastics implicated but also the useful life of plastics as a whole and assure durability at all manufacturing, distribution, utilization, and disposal of waste phases. The development, production, chemical extraction, and end product of biodegradable plastics are all essential elements. This emphasizes the requirement for openness in the manufacture and processing of plastics.

4.3 Waste Recycling and Upcycling Technologies

Waste recycling and waste upcycling are distinct methods for managing waste. Waste recycling is a systematic process that transforms waste materials into new products using mechanical or chemical procedures. This approach requires breaking down the waste material into its fundamental components and reusing them to create new products. In contrast, waste upcycling entails transforming waste materials into higher-value products through innovative means. The process involves taking waste

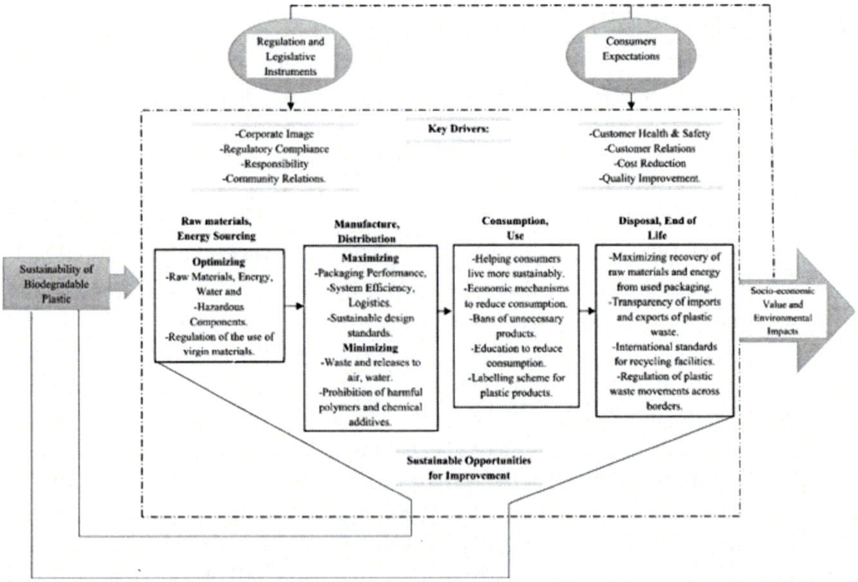

Fig. 4.8 Conceptual framework for sustainability and enhanced policy implementation on biodegradable packaging. Adapted with permission from [22], Elsevier, 2022

materials and repurposing them to create new products. The key disparity between recycling and upcycling is that recycling involves dismantling waste materials into their elementary parts to create new products, while upcycling involves inventively transforming waste materials into new products without entirely dismantling them. Upcycling is commonly perceived as sustainable and eco-friendly waste management approach as it avoids the necessity for energy-intensive recycling processes while adding value to waste materials that would otherwise be discarded. Figure 4.9 shows the market-based recycling strategies and their end products.

4.3.1 *Mechanical Recycling of Plastics*

The method of mechanical recycling includes mechanically melting and grinding plastic trash to generate regrind, which may be utilized to make new items. Because of its affordability and high level of dependability, this technique is regarded as one of the most desired recycling strategies. The fundamental benefit of mechanical recycling is that it preserves the polymer's molecular structure, enabling the development of new products with qualities comparable to that of goods developed from virgin components. Primary and secondary mechanical recycling are the two categories. Although secondary recycling entails recycling used plastic waste, primary mechanical recycling refers to the reuse of plastics waste produced during the production

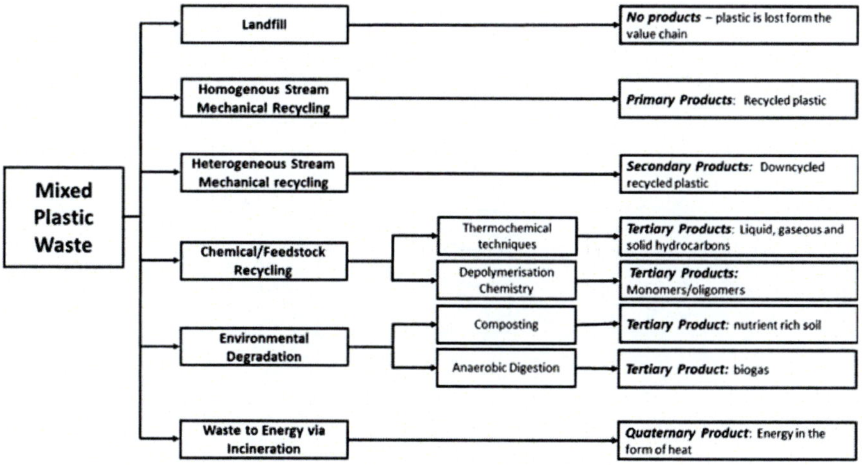

Fig. 4.9 Different types of recycling plastic waste and their product formation. Reproduced with permission from [23], Elsevier, 2021

phase. For the production of new goods by primary mechanical recycling, waste materials must first be collected and sorted before being ground, melted, and extruded, as shown in Fig. 4.10. Post-consumer plastics trash is collected, sorted, and cleaned as part of secondary mechanical recycling, which subsequently processes it similarly to primary garbage [25].

The mechanism of mechanical recycling involves the grinding of plastics waste material into small particles, which are then melted and reprocessed to create new products. During the grinding process, the waste material is reduced in size, increasing the surface area and allowing for better melting and mixing of the material during reprocessing. After grinding, the waste material is melted and extruded into pellets, which can be used to create new products. The process of mechanical recycling involves several steps, including collection, sorting, grinding, melting, and extrusion. The first step is to collect the plastics waste material, which is then sorted based on its type and composition. The sorted waste material is then ground into small particles, which are melted and extruded into pellets. These pellets can be used to create new products, either alone or in combination with virgin materials [26].

4.3.2 Chemical Recycling of Plastics

Unlike mechanical recycling, which relies on the mechanical grinding and melting of plastics waste material, chemical recycling involves breaking down the polymer molecules into their constituent monomers or other useful chemicals through chemical reactions. This approach can be used to recycle a wide range of plastics, including

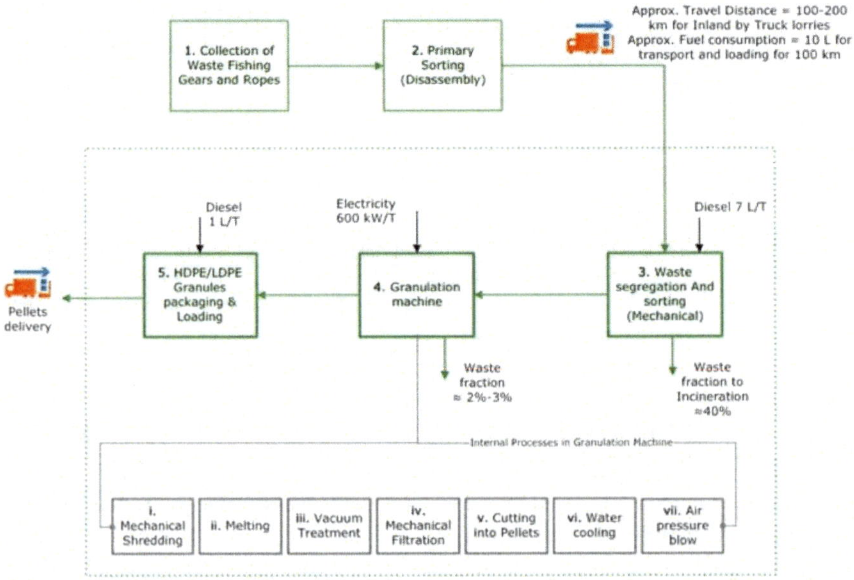

Fig. 4.10 Process of mechanical recycling of plastics, practiced in Norway. Reproduced with permission from [24], Elsevier, 2020

those that are not suitable for mechanical recycling due to their complexity or contamination. The mechanism of chemical recycling involves the use of chemical reactions to break down the polymer molecules into their constituent monomers or other useful chemicals. There are several methods of chemical recycling, including pyrolysis, depolymerization, and gasification, Fig. 4.11. Pyrolysis involves the heating of the plastics waste material in the absence of oxygen, resulting in the breakdown of the polymer molecules into smaller fragments. Depolymerization, on the other hand, involves the use of chemicals to break down the polymer molecules into their constituent monomers. Gasification involves the conversion of the plastics waste material into a mixture of gases, which can be further processed to create useful chemicals or fuels.

Despite its advantages, chemical recycling faces several challenges. One of the main challenges is the high cost and energy consumption associated with the process, particularly in the case of pyrolysis and gasification. In addition, the quality of the recycled products may be lower than that of virgin materials, which can limit their use in certain applications. Furthermore, the scalability of chemical recycling technologies is still limited, which means that it may take time before the technology can be adopted on a large scale. Pyrolysis is suitable for plastic waste that cannot be recycled mechanically like PVC, mixed plastics, although it is energy-intensive and may result in greenhouse gas emissions. While hydrocracking and polymerization are effective for creating new plastics, they necessitate substantial investment in infrastructure and technology. Gasification can generate a beneficial gas, but it

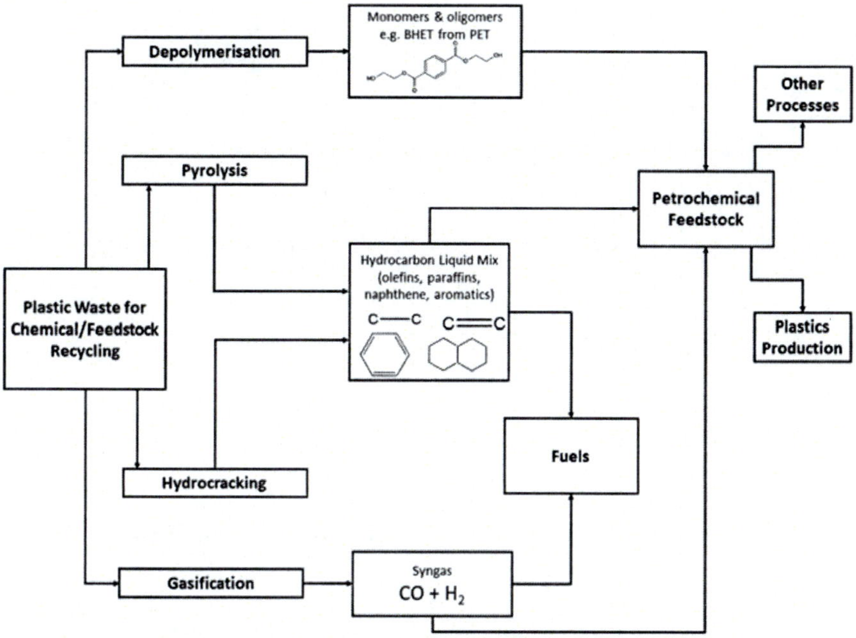

Fig. 4.11 Types and process of chemical recycling. Reproduced with permission from [24], Elsevier, 2020

requires careful management of waste streams and may lead to emissions. However, ongoing research and development efforts are aimed at addressing these challenges and improving the efficiency and sustainability of chemical recycling [27].

4.3.3 Microwave-Assisted Plastic Conversion

Microwave-assisted plastic conversion, methodology as schematically shown in Fig. 4.12, is a promising method for converting plastic waste into valuable chemicals and fuel. The process involves the use of microwave radiation to break down the long polymer chains of plastics into smaller molecules, such as hydrocarbons, via a process called pyrolysis. There are two main methods of microwave-assisted pyrolysis: batch and continuous pyrolysis. In batch pyrolysis, plastic waste is heated in a sealed reactor using microwave radiation until it breaks down into smaller molecules. The resulting gases are then condensed and collected. In continuous pyrolysis, plastic waste is fed into a continuous reactor where microwave radiation is used to heat the waste as it flows through the reactor. The resulting gases are continuously collected and condensed.

However, there are several challenges associated with microwave-assisted plastic conversion. One of the main challenges is the heterogeneity of plastic waste, as

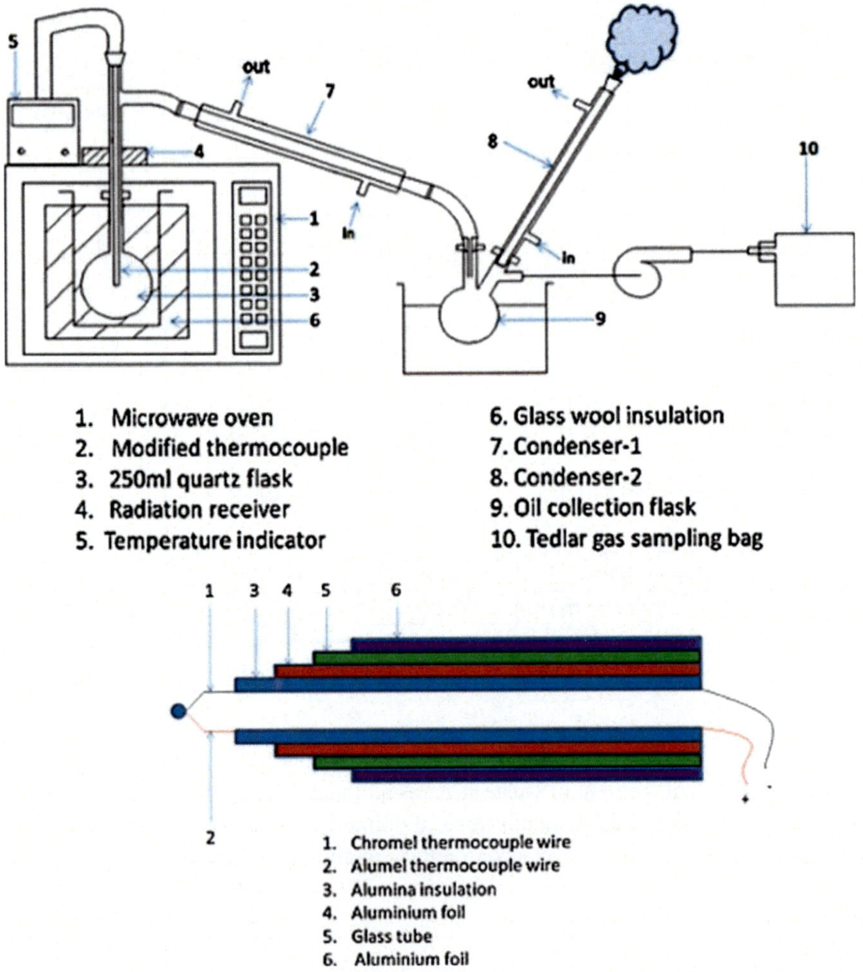

Fig. 4.12 Microwave pyrolysis experimental setup and the thermocouple construction. Reproduced with permission from [28], Elsevier, 2022

different types of plastic require different temperatures and reaction times for pyrolysis due to their differing chemical structures. This can make it challenging to optimize the process for maximum yield and selectivity. Another challenge is the design of the reactor used for the process. Microwave-assisted pyrolysis requires a specially designed reactor that can withstand high temperatures and pressures, and the reactor's design is critical to achieving high yields and selectivity. Energy efficiency is also a concern for this process, as microwave radiation requires a lot of energy, which can make the process less economically viable. Therefore, finding ways to increase the energy efficiency of the process is essential to make it sustainable [29].

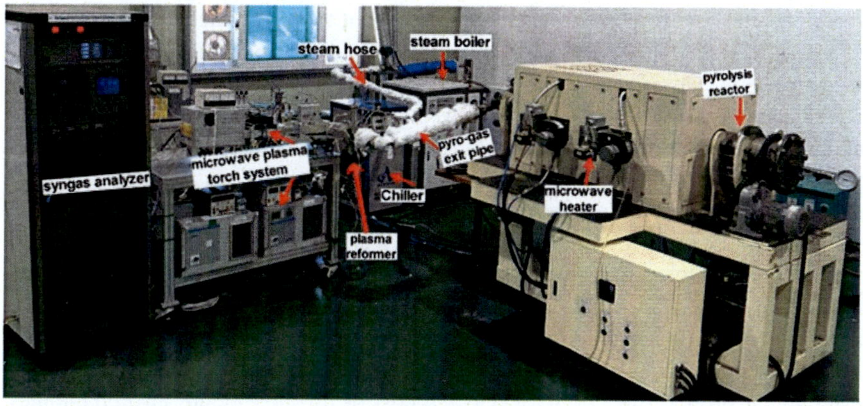

Fig. 4.13 The pilot-scale plant for pure H_2 production via plasma reformer using plastic resources, capable of processing 100 kg/day. Reproduced with permission from [30] Elsevier, 2023

4.3.4 Plasma-Assisted Conversion and Supercritical Conversion

Plasma-assisted conversion of plastic and supercritical water conversion of plastic are two promising processes for converting plastic waste into useful chemicals and fuel. Plasma technology is used to break down plastic waste into smaller molecules by subjecting it to high-energy plasma. The resulting gases and liquids can be used as feedstock for chemical and fuel production. Two methods of plasma-assisted conversion of plastic are direct plasma treatment and plasma-assisted gasification. The conversion efficiency and product yield depend on factors such as plasma power and gas flow rate, plastic waste properties, and residence time (Fig. 4.13).

Supercritical water conversion of plastic is a process that uses water in its supercritical state to break down plastic waste into smaller molecules. The process involves exposing plastic waste to supercritical water, which can be done through batch or continuous treatment methods. The conversion efficiency and product yield depend on factors such as temperature and pressure, residence time, and water quality. However, challenges such as corrosion and erosion of the reactor and difficulty in separating and purifying the products need to be addressed to make the process more sustainable [31].

4.3.5 Emerging Techniques

Photoreforming and compatibilization are two promising methods for recycling plastic waste. Photoreforming uses light to break down plastic waste into smaller chemical building blocks. This process is performed in the presence of a catalyst that accelerates the breakdown of plastic waste without requiring high temperatures or pressures, reducing energy consumption and costs. Photoreforming is effective for a wide range of plastic waste types, including mixed plastics that are challenging to recycle using traditional methods. However, the development of effective catalysts and the cost of technology are still challenges to be addressed. Compatibilization, on the other hand, involves adding additives to different types of plastics to make them compatible with each other, thereby expanding the range of materials that can be recycled. This process modifies the surface properties of plastics, allowing them to mix and bind together using additives, such as maleic anhydride. The process enables the recycling of mixed plastics, which are also challenging to recycle using traditional methods. Compatibilization can reduce plastic waste that ends up in landfills or the environment and increase the strength and durability of recycled plastics. The development of effective additives and the cost of technology are also challenges that need to be addressed [31].

As shown in Fig. 4.14 manufacturing can be divided into additive manufacturing, subtractive manufacturing, and forming.

The additive manufacturing technologies (Fig. 4.14) needs less machining, less assembly, and fewer manufacturing steps, which saves large energy consumption. The raw materials used in the manufacturing process of many additive manufacturing technologies are powders, and many powders are produced during the additive manufacturing process. To achieve the purpose of sustainable materials and a circular economy, the waste powder generated in the additive manufacturing process are collected and recycled for reuse.

The technical flow of waste powder collection, recovery, treatment, and reuse in directed energy deposition (DED) technology in additive manufacturing is shown in Fig. 4.15. After the first energy-directed deposition, a lot of waste powder is

Fig. 4.14 Different types of manufacturing technology

generated, which is collected on the deposition platform. After finishing the collected waste powder, the material is characterized by a scanning electron microscope, and the particle size distribution (PSD) of the collected waste powder is analyzed [32]. If the particle size of the waste powder is smaller than the required size, you can directly proceed to the next step; otherwise, the waste powder needs to be screened to screen out the powder particles that are too large. Then, the qualified powder after screening is mixed with new powder, and a part of the mixed powder particles are screened out for material characterization analysis. If the performance of these powder particles meets the standard, the mixed powder will be recycled and can be reused. These powders are added to the powder storage plate for the next deposition, and subsequent waste powder collection and reuse can recycle the method process [32].

Powder-based fused (PBF) is another additive manufacturing technology. Recent research technique of precipitated polybutylene terephthalate feedstock material for powder bed fusion of polymers is shown in Fig. 4.16. The new powder is powder-based fused (selective laser melting), the remaining waste powder is screened, and the missing powder that can no longer be used is screened out. The screened powder can be used for the next time PBF after a powder performance analysis and screening [33].

In these two waste powder collection methods, there is still a lot of waste powder that is sieved out because the size does not meet the standard or the powder performance is not good, and these powders are not reused. However, follow-up research can be carried out to propose a treatment method for these waste powders so that they can meet the required standards for reuse. In this way, the recovery rate and reuse rate of waste powder are greatly improved, reflecting the trend of material sustainability and circular economy [33].

3D printing is a rapidly growing technology that generates a significant amount of plastic waste, making 3D printed upcycling an essential process for a more sustainable manufacturing industry. This process involves collecting and processing plastic waste generated by 3D printing and using it to create new 3D printed objects. However, challenges such as developing effective recycling methods that can efficiently process

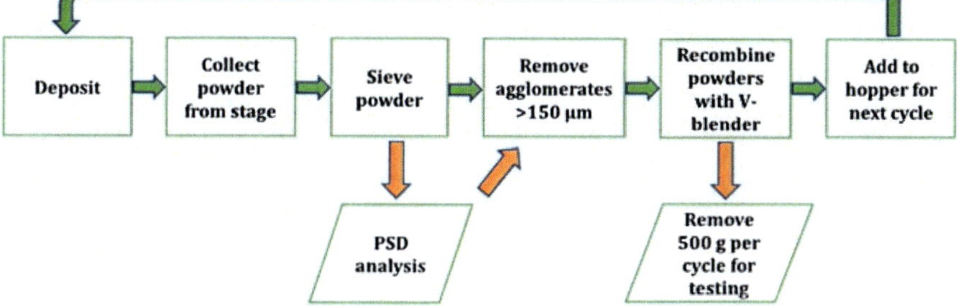

Fig. 4.15 Waste collection technologies for DED powder

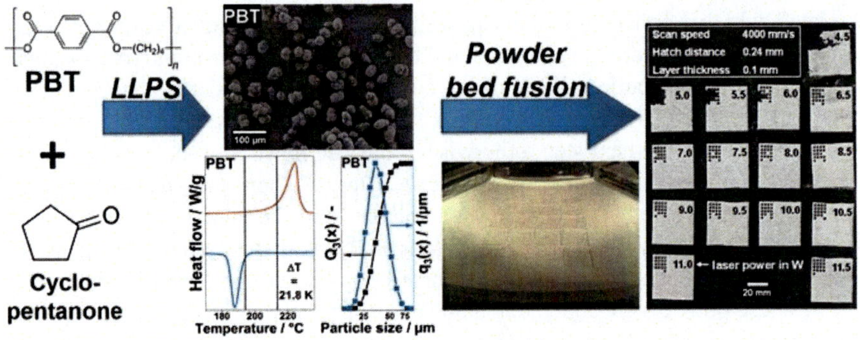

Fig. 4.16 Flow diagram of development of feedstock material for PBF by precipitated polybutylene terephthalate material and initial process ability in PBF [33]

and purify plastic waste and the potential lower quality of recycled plastic compared to virgin plastic limit its suitability for some applications.

4.3.6 *Recycling Techniques for PET/HDPE*

This section will compare the current technologies used for PET, HDPE and lightweight packaging with emerging technologies coming up or being researched on. Many countries are currently sharing the same existing methods of processing some of these materials, therefore this section will explore other countries (apart from Canada and Korea) to discuss one emerging technology.

This section will only focus on post-consumer recycled PET. There are two main methods to recycle and process post-consumer PET, Mechanical and Chemical. With a new emerging method called "Cure technology" it will simplify and reduce the worries of contamination in the mechanical process. Mechanical processing involves using PET flakes in the contamination removal process, washing, drying and melting. During the contamination removal stage, PET bottles need to be separated from other plastic because contamination from other plastics will cause an even more significant deterioration of post-consumer PET. These bottles will then be grounded into flakes. For PVC/PET, PVC can be recognized using an automatic separation technique based on the detection of chlorine atoms. According to the findings, the multistage grinding process known as the micronyl procedure successfully eliminated 97.5% of PVC. Post-consumer PET is sorted, then ground into flakes so it can be easily recycled. The PET flakes are then rinsed after grinding [13]. The PET flakes can be cleaned using one of two techniques: (1) washing with hot water and a NaOH solution followed by cold water, (2) tetrachloroethylene usage (TCE). The process of drying has emerged as being essential to Postconsumer PET recycling. Reducing the moisture level of the post-consumer PET flakes lessens the influence of hydrolytic degradation and improves Recycled PET melt strength. The majority of Postconsumer PET producers employ drying conditions of 3–7 h at 140–170 °C. The maximum amount of water that is permitted to be present inside PET flakes in ordinary working conditions is 50 ppm; this is typically achieved with desiccated dryers operating at 170 °C for 6 h prior to extrusion. Polyesters may be recycled by breaking the polymer chains down into their unique monomers or different low molecular weight monomers that may be eventually re-polymerized into new PET. A substantial gain of this method is the ability to purify the monomer, disposing of stable particulates, which include pigments, and chemical impurities, which include residual catalysts or dyes. However, there is a downside to greenhouse emissions energy consumption. This is the basis of the new technology called "cure technology". Cure technology (see Fig. 4.17) is a new technology developed in the Netherlands (and jointly tested in Canada) that uses partial depolymerization. The objective of Cure is to decolorize all varieties of post-consumer polyester and transform them into clear pellets with qualities identical to those of virgin polyester, making them acceptable for demanding applications including carpets, textiles, and food packaging. A fully circular polyester chain cannot be achieved until 100% post-consumer polyester is recycled using low energy [34].

HDPE recycling is nearly identical to PET, with the most crucial distinction in how they may be sorted [14]. They are separated primarily based totally on their thickness because it differs primarily based totally on the device used. After those preliminary stages, the plastic can also go through homogenization if it is not always constructed

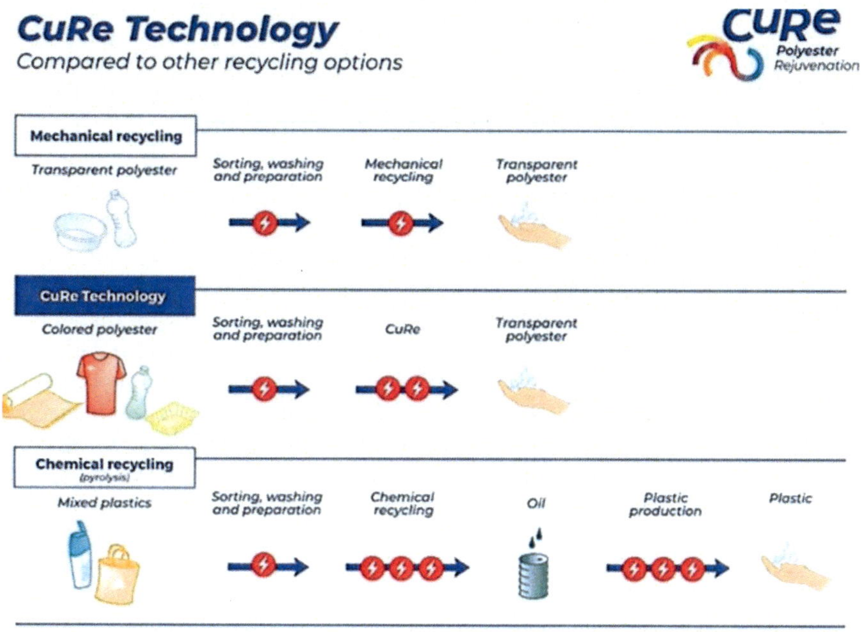

Fig. 4.17 Concepts involved in cure technology [34]

from HDPE. Homogenization separates the HDPE merchandise and portions so that some other plastics they had been blended with no longer inhibit HDPE-unique recycling. Recycling agencies can also additionally perform homogenization with numerous methods. For instance, they can isolate PET plastic from HDPE via sink-flow separation, in which the distinct densities of those substances can have them flow at distinct ranges in a liquid. They may also differentiate among HDPE and different plastic gadgets by hitting them with infrared radiation and locating their precise near-infrared (NIR) signatures. It undergoes granulation with the HDPE nicely eliminated from different plastics and debris. Here, HDPE is first shredded and softened by machines before being reformed into homogeneous granules. These pellets serve as the basic building components of recycled goods. An employer must combine several pellets by blending them at high temperatures and moulding them into a novel shape to produce a product like plastic lumber from them [14]. One of the main issues of recycle HDPE (HDPE) that it has poorer mechanical strength as compared to virgin HDPE. One of the significant problems with recycled HDPE (rHDPE) is its lower mechanical strength compared to virgin HDPE. To solve this, various rHDPE sources are blended with PE100-grade raw HDPE in various ratios [35]. The blend is completely characterized to evaluate whether pipe applications are feasible. All blends have characteristics that are greater than the minimal standards necessary for PE100 grades, including tensile strength at yield, elongation at break, and flexural modulus. Additionally, resistance to slow crack growth (SCG) and rapid fracture propagation (RCP), two crucial mechanical characteristics of polyethylene pipes, are thoroughly assessed. Unexpectedly, a twofold correlation between SCG and RCP as well as recycled PE content in blends was found, allowing the creation of predictive capabilities to guarantee pressure pipe application needs and specifications [36]. Sample data of the mechanical strength can be seen in Lightweight Packaging Processing Technologies A typical process flow diagram of a treatment facility for processing light packaging, commercial garbage, and domestic waste is shown in Fig. 27. Using a counter comb shredder, the material is released after loading and sent to the screening stage. The screen step has two purposes. (1) Verifying that fine grain size items are being deposited; and (2) preparing coarser sieving residues for classification and sorting [37].

A continual supply of the overflow product stream occurs in order to function at high efficiency. Additionally, it's important to avoid both under- and overfilling the future process phases. The screening procedure is a necessary step for good separation, and subsequent processing procedures (such as air separation and magnetic separation) call for the material to have a monolayer morphology in order to obtain a high level of separation [38]. However, there are some common issues faced with this current process. (i) Municipal waste, mineral products, and post-consumer waste such light packaging does not have mass characteristics and cannot flow or be collected. Gravimetric weighing techniques, such as belt scales, are inappropriate for controlled material flow management because plastic raw materials have a low density (ii) The waste combinations are quite diverse; thus the capacitive design of the devices must handle varying amounts of material. The machine must function consistently efficiently over the designated range. Underfilling and overfilling ("peaks"), however, cannot be completely ruled out (iii) In the case of municipal garbage and light packing, which is often a drum screen, the parameterization of the first separation unit (inclination and speed) is adjusted to mean the residence time for a certain material flow. It is unable to react to changes in flow rate or material composition because these parameters are fixed (iv) The fundamental issue with sorting or processing facilities is that, despite the facility operating throughout the day, there is not a constant supply of material to be processed. Wheel loaders, a type of mobile loading device, are frequently employed for loading. Mobile loading technology is not always accessible for system loading due to multitasking (for instance, delivery and loading of processed products) and waste treatment facilities. Controlling the material flow is the main goal in order to provide a constant flow of materials. This aims to establish ideal conditions for categorization and sorting procedures. The following are the concept's control variables: (1) The speed of the V-belt, which connects the anti-comb crusher and the drum screen, in relation to the volume flow that was observed; (2) The truck's anti-tendency pillar's to crash; (3) The process of integrating mobile technology into sewage treatment plants using optical warning messages (traffic lights) To ensure the circumstances of good sequencing of the sorting through a continuous flow of the volume of material and to prevent additional filling and underfilling, the necessary condition of the material flow for the succeeding stages of sorting must be enhanced. Results from the mechanical portion of the mechanical–biological waste treatment facility demonstrated that guided feeding could increase feed efficiency. The tasks can be accomplished through volume-dependent sensor control of the conveyor belt's speed, taking into consideration the anti-comb unit's loading behaviour and traffic light signals for loading employees [16, 39].

4.4 ESG in Plastic Waste

ESG factors are crucial in managing plastic waste due to its severe environmental and social impact, which harm wildlife, contaminate water sources, and damage ecosystems. Marginalized communities and developing countries, which lack waste management infrastructure, are disproportionately affected. Companies that prioritize ESG considerations, including reducing plastic waste, adopting a circular economy approach, engaging with local communities, supporting waste management infrastructure, ensuring compliance with environmental regulations, establishing internal policies, and being transparent, are more likely to be sustainable.

4.4.1 Extended Producer Responsibility (EPR)

Extended Producer Responsibility (EPR) is a policy approach that holds producers responsible for the end-of-life management of their products, including the cost of recycling or disposing of them. The EU has had an EPR policy for packaging waste since 1994, which requires producers to finance and organize the collection and recycling of their packaging waste. In 2012, the EU introduced an EPR policy for electronic waste, requiring producers to finance the collection and recycling of electronic waste [40]. As of 2021, 21 EU member states have implemented EPR policies for packaging waste, and all EU member states have implemented EPR policies for electronic waste. According to the European Environment Agency, in 2018, the average collection rate for packaging waste in the EU was 67.3%, and the average recycling rate was 44.9% [41]. One of the main challenges faced by the EU in implementing EPR policies has been ensuring compliance with the regulations, as well as reducing the administrative burden for businesses. Additionally, there are concerns about the effectiveness of EPR policies in reducing waste, as they may not address the root causes of waste generation [42].

Producers have several responsibilities under Extended Producer Responsibility (EPR) programs. Firstly, they are legally liable for any environmental damage caused by their products during their entire life cycle. They must bear the costs of collecting, recycling or disposing of their products after use, and are responsible for designing products that are easier to recycle or dispose of [43, 44]. Producers retain ownership of their products throughout their life cycle and must provide consumers with information about the environmental impact of their products. It was recommended that consumers engage in several activities. Firstly, they were advised to sort plastic waste from other types of waste and categorize it based on its type and recyclability [45]. Secondly, it was suggested that they use designated recycling bins for plastic waste, either at home or in public places. Thirdly, consumers were reminded to avoid contaminating plastic waste with other types of waste, such as food waste or hazardous materials. Finally, it was recommended that consumers opt for sustainable alternatives such as reusable bags and bamboo bottles to reduce plastic consumption, thereby minimizing the amount of single-use plastics that ended up in landfills.

4.4.2 Policies and Schemes

Plastic waste management has been a significant environmental concern globally. In response, various cities and countries have implemented plastic bag bans or fees. California became the first US state to ban single-use plastic bags in 2014 [46]. Other countries, such as Ireland and Kenya, have also implemented plastic bag bans or fees. These policies have effectively reduced plastic waste generation. For instance, after implementing a plastic bag ban in 2007, plastic bag usage in San Francisco dropped by over 70% within the first year [47]. Deposit and return schemes (DRS) have also proven successful in reducing plastic waste. In Germany, where a DRS has been in place for over a decade, the return rate for plastic bottles is over 90%. Consumers can receive a refund for returning used plastic bottles to designated recycling centers, incentivizing them to recycle plastic bottles and helping to reduce the amount of plastic waste that ends up in landfills or oceans [48, 49]. Governments enforced regulations that require manufacturers to design products that are more environmentally friendly and use less plastic. Effective plastic waste collection and recycling programs are essential for proper plastic waste management. Governments have established curbside recycling programs, drop-off recycling centers, and public education campaigns to promote recycling. In 2018, the national recycling rate for plastics in the United States was only 8.7% [50]. However, with proper waste collection and recycling programs, plastic waste can be diverted from landfills and oceans. Plastic-to-fuel technology is a promising policy that can reduce the amount of plastic waste that ends up in landfills and oceans. This technology converts plastic waste into fuel, providing an alternative to fossil fuels. Japan, the United States, and India are among the countries that have implemented this policy, and its use is growing globally [51, 52]. International agreements, such as the Basel Convention, help regulate the movement of plastic waste across borders and encourage countries to reduce their plastic waste generation. The United Nations has set a Sustainable Development Goal (SDG) to significantly reduce marine pollution by 2025, emphasizing the importance of international cooperation in addressing the plastic waste crisis [53].

4.5 Case Studies

In the given example, the LCA analysis of HDPE was conducted using Open LCA software, which is a widely used software for LCA analysis. The analysis was conducted on the production and end-of-life stages of cutting in HDPE [35], which is a commonly used application of HDPE. For the analysis, the preloaded Agribalyse project was used, which is a comprehensive life cycle inventory database containing data on various products and their environmental impacts. The database includes data on the production and end-of-life stages of HDPE cutting, which were used for the analysis. The geographic location of Ontario, Canada was selected for the analysis. This is important because the environmental impacts of a product can vary based on the location where it is produced and used. Method included in openLCA LCIA method package 2.1.1, compatible with ecoinvent v3.6, v3.7, ReCiPe 2016 Endpoint (H) [36].

The goal of this LCA analysis is to determine the environmental impacts of HDPE (High-Density Polyethylene) production and disposal in Canada. The LCA methodology used in this study follows the ISO 14040 series of standards. The LCA will consider three stages: (1) plastic production, (2) plastic use, and (3) plastic disposal. The data used in the study will be collected from primary sources and secondary sources such as literature reviews, databases, and industry reports. The the results is presented in terms of environmental impacts per unit of plastic produced. The functional unit is 10 kg of HDPE resin (Table 4.3).

Table 4.3 Impact assessment parameters

Name	Reference unit
Fine particulate matter formation	DALY
Fossil resource scarcity	USD2013
Freshwater ecotoxicity	species.yr
Freshwater eutrophication	species.yr
Global warming, Freshwater ecosystems	species.yr
Global warming, Human health	DALY
Global warming, Terrestrial ecosystems	species.yr
Human carcinogenic toxicity	DALY
Human non-carcinogenic toxicity	DALY
Ionizing radiation	DALY
Land use	species.yr
Marine ecotoxicity	species.yr
Marine eutrophication	species.yr
Mineral resource scarcity	USD2013
Ozone formation, Human health	DALY
Ozone formation, Terrestrial ecosystems	species.yr
Stratospheric ozone depletion	DALY
Terrestrial acidification	species.yr
Terrestrial ecotoxicity	species.yr
Water consumption, Aquatic ecosystems	species.yr
Water consumption, Human health	DALY
Water consumption, Terrestrial ecosystem	species.yr

4.5.1 Interpretation of Results

The findings of the LCA analysis indicate that plastic production and disposal have a significant impact on the environment. The impact categories with the highest impact values were global warming, freshwater ecosystems, and marine ecotoxicity. This suggests that plastic production and disposal are major contributors to climate change and negatively impact aquatic ecosystems. These parameters include reducing fossil fuel consumption, minimizing water usage, and improving waste management practices. By focusing on these parameters, it is possible to reduce the environmental impact of plastic production and disposal. It is important to note that the impact values reported in the analysis are based on a set of assumptions and data inputs and may not be representative of all plastic products and disposal scenarios (Table 4.4).

Table 4.4 Results of impact assessment

Environmental parameter	Reference unit	Impact category	Impact value
Fine particulate matter formation	DALY	Human health	7.2E−06
Fossil resource scarcity	USD2013	Resource depletion	3.5E+01
Freshwater ecotoxicity	species.yr	Ecotoxicity	4.6E−05
Freshwater eutrophication	species.yr	Eutrophication	6.2E−04
Global warming, freshwater ecosystems	species.yr	Climate change	6.8E−04
Global warming, human health	DALY	Human health	1.1E−02
Global warming, terrestrial ecosystems	species.yr	Climate change	1.6E−03
Human carcinogenic toxicity	DALY	Human health	2.3E−05
Human non-carcinogenic toxicity	DALY	Human health	7.1E−07
Ionizing radiation	DALY	Human health	5.5E−07
Land use	species.yr	Land use	1.1E−01
Marine ecotoxicity	species.yr	Ecotoxicity	3.3E−03
Marine eutrophication	species.yr	Eutrophication	1.1E−04
Mineral resource scarcity	USD2013	Resource depletion	3.3E−01
Ozone formation, human health	DALY	Human health	5.6E−06
Ozone formation, terrestrial ecosystems	species.yr	Climate change	4.4E−06
Stratospheric ozone depletion	DALY	Human health	7.7E−08
Terrestrial acidification	species.yr	Acidification	1.8E−05
Terrestrial ecotoxicity	species.yr	Ecotoxicity	3.9E−05
Water consumption, aquatic ecosystems	species.yr	Water use	5.6E−03
Water consumption, human health	DALY	Human health	4.7E−05
Water consumption, terrestrial ecosystem	species.yr	Water use	4.4E−04

Note DALY refers to Disability-Adjusted Life Years, and USD2013 refers to the monetary value of resources based on the year 2013

4.6 Conclusion

In conclusion, this chapter has highlighted the paradox of plastic, where its value is in direct conflict with its lifespan. The principles of circularity in plastics, moisture control, and ash and carbon content were discussed in detail. Additionally, the various end-of-life options such as landfill, incineration, and composting were analyzed. Waste recycling and upcycling technologies, including mechanical and chemical recycling, microwave-assisted and plasma-assisted conversion, and supercritical conversion, were also examined. Furthermore, ESG considerations in plastic waste, such as extended producer responsibility (EPR) and policies and schemes, were discussed. Finally, case studies were presented, and the chapter concludes that an integrated approach utilizing a combination of these technologies, policies, and schemes can help achieve sustainability in the plastic industry.

Activity 4.0—Multiple Choice Questions

1. Which of the following is a primary circularity principle for plastics?
   a. Reduce
   b. Reuse
   c. Recycle
   d. All of the above
2. What is the most commonly used end-of-life option for plastic waste?
   a. Landfill
   b. Incineration
   c. Composting
   d. Recycling
3. Which of the following is a mechanical recycling technique for plastics?
   a. Pyrolysis
   b. Gasification
   c. Extrusion
   d. Depolymerization
4. Which of the following is a chemical recycling technique for plastics?
   a. Incineration
   b. Composting
   c. Depolymerization
   d. Landfill
5. Which of the following recycling techniques uses microwaves to convert plastic waste?
   a. Mechanical recycling
   b. Chemical recycling
   c. Microwave-assisted conversion
   d. Supercritical conversion
6. Which of the following esg schemes is designed to encourage companies to take responsibility for the disposal of their products?
   a. Extended producer responsibility (EPR)
   b. Carbon offsetting
   c. Renewable energy certificates (RECs)
   d. LEED certification
7. What is the primary objective of EPR?
   a. To reduce the amount of waste generated
   b. To improve the efficiency of waste management systems

c. To ensure that companies take responsibility for the disposal of their products
d. To encourage companies to use more sustainable materials

8. Which of the following is an example of an ESG Policy aimed at reducing plastic waste?

   a. Plastic Bag Ban
   b. Carbon tax
   c. Renewable energy mandate
   d. Net-zero emissions target

9. Which of the Following plastics is commonly used in beverage bottles?

   a. Polyethylene terephthalate (PET)
   b. High-density polyethylene (HDPE)
   c. Polyvinyl chloride (PVC)
   d. Low-density polyethylene (LDPE)

10. Which of the following is a common source of plastic pollution in oceans?

    a. Microplastics
    b. Single-use plastic bags
    c. Plastic straws
    d. All of the above

11. What is the primary benefit of upcycling plastic waste?

    a. Reducing waste
    b. Reducing greenhouse gas emissions
    c. Reducing the need for virgin materials
    d. All of the above

12. Which of the following recycling techniques uses plasma to convert plastic waste?

    a. Mechanical recycling
    b. Chemical recycling
    c. Plasma-assisted conversion
    d. Supercritical conversion

13. What is the primary objective of the circular economy?

    a. To reduce waste and improve resource efficiency
    b. To reduce greenhouse gas emissions
    c. To promote sustainable development
    d. To achieve zero waste

14. Which of the following materials is commonly used as a feedstock for chemical recycling of plastics?

    a. Natural gas

   b. Coal
   c. Biomass
   d. All of the above

15. Which of the following is an example of an ESG policy aimed at reducing plastic waste in oceans?
   a. The Ocean Cleanup
   b. The Great Pacific Garbage Patch
   c. The Clean Ocean Act
   d. The International Convention for the Prevention of Pollution from Ships

16. Which of the following is an example of an upcycling technology for plastic waste?
   a. Gasification
   b. Pyrolysis
   c. Depolymerization
   d. 3D printing

17. Which of the following recycling techniques is considered to be the most environmentally friendly?
   a. Mechanical recycling
   b. Chemical recycling
   c. Landfill
   d. Incineration

Solutions

1. d. All of the above
2. a. Landfill
3. c. Extrusion
4. c. Depolymerization
5. c. Microwave-assisted conversion
6. a. Extended producer responsibility (EPR)
7. c. To ensure that companies take responsibility for the disposal of their products
8. a. Plastic Bag Ban
9. a. Polyethylene terephthalate (PET)
10. d. All of the above
11. d. All of the above
12. c. Plasma-assisted conversion
13. a. To reduce waste and improve resource efficiency
14. d. All of the above
15. a. The Ocean Cleanup
16. d. 3D printing
17. a. Mechanical recycling

18. Single-use plastic items such as bags, straws, and bottles.

References

1. Liang Y, Tan Q, Song Q, Li J (2021) An analysis of the plastic waste trade and management in Asia. Waste Management 119:242–253. https://doi.org/10.1016/J.WASMAN.2020.09.049
2. Ncube LK, Ude AU, Ogunmuyiwa EN, et al (2021) An Overview of Plastic Waste Generation and Management in Food Packaging Industries. Recycling 2021, Vol 6, Page 12 6:12. https://doi.org/10.3390/RECYCLING6010012
3. Kuan SH, Low FS, Chieng S (2022) Towards regional cooperation on sustainable plastic recycling: comparative analysis of plastic waste recycling policies and legislations in Japan and Malaysia. Clean Technologies and Environmental Policy 2021 24:3 24:761–777. https://doi.org/10.1007/S10098-021-02263-0
4. Hahladakis JN, Purnell P, Iacovidou E, et al (2018) Post-consumer plastic packaging waste in England: Assessing the yield of multiple collection-recycling schemes. Waste Management 75:149–159. https://doi.org/10.1016/J.WASMAN.2018.02.009
5. Antonopoulos I, Faraca G, Tonini D (2021) Recycling of post-consumer plastic packaging waste in the EU: Recovery rates, material flows, and barriers. Waste Management 126:694–705. https://doi.org/10.1016/J.WASMAN.2021.04.002
6. Hu Q, Ok YS, Wang CH (2022) Sustainable and Highly Efficient Recycling of Plastic Waste into Syngas via a Chemical Looping Scheme. Environ Sci Technol 56:8953–8963. https://doi.org/10.1021/ACS.EST.2C01645/SUPPL_FILE/ES2C01645_SI_001.PDF
7. Kleinhans K, Demets R, Dewulf J, et al (2021) Non-household end-use plastics: the 'forgotten' plastics for the circular economy. Curr Opin Chem Eng 32:100680. https://doi.org/10.1016/J.COCHE.2021.100680
8. Lopez G, Artetxe M, Amutio M, et al (2018) Recent advances in the gasification of waste plastics. A critical overview. Renewable and Sustainable Energy Reviews 82:576–596. https://doi.org/10.1016/J.RSER.2017.09.032
9. Anuar Sharuddin SD, Abnisa F, Wan Daud WMA, Aroua MK (2017) Energy recovery from pyrolysis of plastic waste: Study on non-recycled plastics (NRP) data as the real measure of plastic waste. Energy Convers Manag 148:925–934. https://doi.org/10.1016/J.ENCONMAN.2017.06.046
10. Smith RL, Takkellapati S, Riegerix RC (2022) Recycling of Plastics in the United States: Plastic Material Flows and Polyethylene Terephthalate (PET) Recycling Processes. ACS Sustain Chem Eng 10:2084–2096. https://doi.org/10.1021/ACSSUSCHEMENG.1C06845/SUPPL_FILE/SC1C06845_SI_002.XLSX
11. Caprai V, Schollbach K, Florea MVA, Brouwers HJH (2020) Investigation of the hydrothermal treatment for maximizing the MSWI bottom ash content in fine lightweight aggregates. Constr Build Mater 230:116947. https://doi.org/10.1016/J.CONBUILDMAT.2019.116947
12. Zhang J, Liu B, Zhao S, et al (2020) Preparation and characterization of glass ceramic foams based on municipal solid waste incineration ashes using secondary aluminum ash as foaming agent. Constr Build Mater 262:120781. https://doi.org/10.1016/J.CONBUILDMAT.2020.120781
13. Chidepatil A, Bindra P, Kulkarni D, et al (2020) From Trash to Cash: How Blockchain and Multi-Sensor-Driven Artificial Intelligence Can Transform Circular Economy of Plastic Waste? Administrative Sciences 2020, Vol 10, Page 23 10:23. https://doi.org/10.3390/ADMSCI10020023
14. Techawinyutham L, Tengsuthiwat J, Srisuk R, et al (2021) Recycled LDPE/PETG blends and HDPE/PETG blends: mechanical, thermal, and rheological properties. Journal of Materials Research and Technology 15:2445–2458. https://doi.org/10.1016/J.JMRT.2021.09.052

15. Kawecki D, Wu Q, Gonçalves JSV, Nowack B (2021) Polymer-specific dynamic probabilistic material flow analysis of seven polymers in Europe from 1950 to 2016. Resour Conserv Recycl 173:105733. https://doi.org/10.1016/J.RESCONREC.2021.105733
16. Trenson L, Trenson S, Van Nes F, et al (2022) Liraglutide for Weight Management in the Real World: Significant Weight Loss Even if the Maximal Daily Dose Is Not Achieved. Obes Facts 15:83–89. https://doi.org/10.1159/000520217
17. Demetrious A, Crossin E (2019) Life cycle assessment of paper and plastic packaging waste in landfill, incineration, and gasification-pyrolysis. Journal of Material Cycles and Waste Management 2019 21:4 21:850–860. https://doi.org/10.1007/S10163-019-00842-4
18. Yan K, Gao F, Sun H, et al (2019) Effects of municipal solid waste incineration fly ash on the characterization of cement-stabilized macadam. Constr Build Mater 207:181–189. https://doi.org/10.1016/J.CONBUILDMAT.2019.02.048
19. Dedoussi IC, Eastham SD, Monier E, Barrett SRH (2020) Premature mortality related to United States cross-state air pollution. Nature 2020 578:7794 578:261–265. https://doi.org/10.1038/s41586-020-1983-8
20. Zhang Y, Gautam R, Pandey S, et al (2020) Quantifying methane emissions from the largest oil-producing basin in the United States from space. Sci Adv 6:. https://doi.org/10.1126/SCIADV.AAZ5120/SUPPL_FILE/AAZ5120_SM.PDF
21. Sintim HY, Bary AI, Hayes DG, et al (2019) Release of micro- and nanoparticles from biodegradable plastic during in situ composting. Science of The Total Environment 675:686–693. https://doi.org/10.1016/J.SCITOTENV.2019.04.179
22. Moshood TD, Nawanir G, Mahmud F, et al (2022) Sustainability of biodegradable plastics: New problem or solution to solve the global plastic pollution? Current Research in Green and Sustainable Chemistry 5:100273. https://doi.org/10.1016/J.CRGSC.2022.100273
23. Davidson MG, Furlong RA, McManus MC (2021) Developments in the life cycle assessment of chemical recycling of plastic waste – A review. J Clean Prod 293:126163. https://doi.org/10.1016/J.JCLEPRO.2021.126163
24. Deshpande PC, Skaar C, Brattebø H, Fet AM (2020) Multi-criteria decision analysis (MCDA) method for assessing the sustainability of end-of-life alternatives for waste plastics: A case study of Norway. Science of The Total Environment 719:137353. https://doi.org/10.1016/J.SCITOTENV.2020.137353
25. Jeswani H, Krüger C, Russ M, et al (2021) Life cycle environmental impacts of chemical recycling via pyrolysis of mixed plastic waste in comparison with mechanical recycling and energy recovery. Science of The Total Environment 769:144483. https://doi.org/10.1016/J.SCITOTENV.2020.144483
26. Trossaert L, De Vel M, Cardon L, Edeleva M (2022) Lifting the Sustainability of Modified Pet-Based Multilayer Packaging Material with Enhanced Mechanical Recycling Potential and Processing. Polymers 2022, Vol 14, Page 196 14:196. https://doi.org/10.3390/POLYM14010196
27. Chu M, Liu Y, Lou X, et al (2022) Rational Design of Chemical Catalysis for Plastic Recycling. ACS Catal 12:4659–4679. https://doi.org/10.1021/ACSCATAL.2C01286/SUPPL_FILE/CS2C01286_SI_001.PDF
28. Ramasubramanian, B., Tan, J., Chellappan, V., & Ramakrishna, S. (2023). Recent Advances in Extended Producer Responsibility Initiatives for Plastic Waste Management in Germany and UK. Materials Circular Economy, 5(1), 6.
29. Fernández-Andrade KJ, Fernández-Andrade AA, Zambrano-Intriago LÁ, et al (2023) Microwave-assisted MOF@biomass layered nanomaterials: Characterization and applications in wastewater treatment. Chemosphere 314:137664. https://doi.org/10.1016/J.CHEMOSPHERE.2022.137664
30. Ganza PE, Lee BJ (2023) A novel method for industrial production of clean hydrogen (H2) from mixed plastic waste. Int J Hydrogen Energy 48:15037–15052. https://doi.org/10.1016/J.IJHYDENE.2023.01.010
31. Maitlo G, Ali I, Maitlo HA, et al (2022) Plastic Waste Recycling, Applications, and Future Prospects for a Sustainable Environment. Sustainability 2022, Vol 14, Page 11637 14:11637. https://doi.org/10.3390/SU141811637

32. Hu YL, Lin X, Li YL, et al (2020) Plastic deformation behavior and dynamic recrystallization of Inconel 625 superalloy fabricated by directed energy deposition. Mater Des 186:108359. https://doi.org/10.1016/J.MATDES.2019.108359
33. Dechet MA, Gómez Bonilla JS, Grünewald M, et al (2021) A novel, precipitated polybutylene terephthalate feedstock material for powder bed fusion of polymers (PBF): Material development and initial PBF processability. Mater Des 197:109265. https://doi.org/10.1016/J.MATDES.2020.109265
34. Welcome at CuRe Technology | Polyester Rejuvenation. https://curetechnology.com/. Accessed 19 Apr 2023
35. Ferreira AM, Sucena I, Otero V, et al (2021) Pretreatment of Plastic Waste: Removal of Colorants from HDPE Using Biosolvents. Molecules 2022, Vol 27, Page 98 27:98. https://doi.org/10.3390/MOLECULES27010098
36. Kulkarni P, Ravekar V, Rama Rao P, et al (2022) Recycling of waste HDPE and PP plastic in preparation of plastic brick and its mechanical properties. Cleaner Materials 5:100113. https://doi.org/10.1016/J.CLEMA.2022.100113
37. Liew JX, Loy ACM, Chin BLF, et al (2021) Synergistic effects of catalytic co-pyrolysis of corn cob and HDPE waste mixtures using weight average global process model. Renew Energy 170:948–963. https://doi.org/10.1016/J.RENENE.2021.02.053
38. Oladele IO, Taiwo AS, Okegbemi TA, et al (2021) Influence of processed waste bagasse fibre-stone dust-6063 aluminium alloy particle on the characteristics of hybrid reinforced recycled HDPE composites 14:909–920. https://doi.org/10.1080/19397038.2021.1923084
39. Zhang X, You S, Tian Y, Li J (2019) Comparison of plastic film, biodegradable paper and bio-based film mulching for summer tomato production: Soil properties, plant growth, fruit yield and fruit quality. Sci Hortic 249:38–48. https://doi.org/10.1016/J.SCIENTA.2019.01.037
40. Pani SK, Pathak AA (2021) Managing plastic packaging waste in emerging economies: The case of EPR in India. J Environ Manage 288:112405. https://doi.org/10.1016/J.JENVMAN.2021.112405
41. Alev I, Agrawal V V., Atasu A (2019) Extended Producer Responsibility for Durable Products. 22:364–382. https://doi.org/10.1287/MSOM.2018.0742
42. Tian F, Sošić G, Debo L (2020) Stable recycling networks under the Extended Producer Responsibility. Eur J Oper Res 287:989–1002. https://doi.org/10.1016/J.EJOR.2020.05.002
43. Bhadra U, Mishra PP (2021) Extended Producer Responsibility in India: Evidence from Recykal, Hyderabad. Journal of Urban Management 10:430–439. https://doi.org/10.1016/J.JUM.2021.07.003
44. Gupt Y, Sahay S (2015) Review of extended producer responsibility: A case study approach. 33:595–611. https://doi.org/10.1177/0734242X15592275
45. Kunz N, Mayers K, Van Wassenhove LN (2018) Stakeholder Views on Extended Producer Responsibility and the Circular Economy. 60:45–70. https://doi.org/10.1177/0008125617752694
46. Atasu A, Subramanian R (2012) Extended Producer Responsibility for E-Waste: Individual or Collective Producer Responsibility? Prod Oper Manag 21:1042–1059. https://doi.org/10.1111/J.1937-5956.2012.01327.X
47. Peng B, Tu Y, Elahi E, Wei G (2018) Extended Producer Responsibility and corporate performance: Effects of environmental regulation and environmental strategy. J Environ Manage 218:181–189. https://doi.org/10.1016/J.JENVMAN.2018.04.068
48. Peng B, Wang Y, Elahi E, Wei G (2019) Behavioral game and simulation analysis of extended producer responsibility system's implementation under environmental regulations. Environmental Science and Pollution Research 2019 26:17 26:17644–17654. https://doi.org/10.1007/S11356-019-05215-W
49. Gu F, Guo J, Hall P, Gu X (2018) An integrated architecture for implementing extended producer responsibility in the context of Industry 4.0. 57:1458–1477. https://doi.org/10.1080/00207543.2018.1489161
50. Lorang S, Yang Z, Zhang H, et al (2022) Achievements and policy trends of extended producer responsibility for plastic packaging waste in Europe. Waste Disposal & Sustainable Energy 2022 4:2 4:91–103. https://doi.org/10.1007/S42768-022-00098-Z

51. Andreasi Bassi S, Boldrin A, Faraca G, Astrup TF (2020) Extended producer responsibility: How to unlock the environmental and economic potential of plastic packaging waste? Resour Conserv Recycl 162:105030. https://doi.org/10.1016/J.RESCONREC.2020.105030
52. Campbell-Johnston K, Calisto Friant M, Thapa K, et al (2020) How circular is your tyre: Experiences with extended producer responsibility from a circular economy perspective. J Clean Prod 270:122042. https://doi.org/10.1016/J.JCLEPRO.2020.122042
53. Richter JL, Koppejan R (2016) Extended producer responsibility for lamps in Nordic countries: best practices and challenges in closing material loops. J Clean Prod 123:167–179. https://doi.org/10.1016/J.JCLEPRO.2015.06.131

Chapter 5
Circular Practices in E-waste Management and Transportation

Abstract This chapter provides a comprehensive overview of electronic waste (e-waste) generation, classification, and recycling strategies. The chapter outlines the various stages of e-waste recycling, including the collection, sorting, dismantling, and advanced recycling techniques. The use of alternate materials and solutions such as the shared economy model, products-as-a-service model, and product ownership model are also discussed. The chapter delves into organic electronics, including the organic field effect transistors, photovoltaics, memory devices, and LEDs. The role of IT-enabled electronics and global initiatives and policies on e-waste management are highlighted. The chapter concludes with case studies that provide insights into various business models and considerations. Overall, this chapter emphasizes the importance of e-waste management and sustainable solutions to mitigate the adverse effects of e-waste on the environment and public health.

Keywords Printed circuit boards · Artificial intelligence · Block chain · Bioleaching

5.1 Overview of Electronic Waste Generation

The phrase "electronic waste" or "e-waste" pertains to discarded electronic devices like computers, televisions, mobile phones, and other digital equipment. This matter has become a mounting global concern, and proper gathering and management are crucial in mitigating its impact on human well-being and the ecosystem. This section will delve into the roots of e-waste, the constituents and parts found in it, and the negative ecological and health effects that can result from improper handling of such waste [1]. Over the course of the last two decades, there has been a swift and substantial acceleration in technological advancement that has led to the development of newer, superior, and more advanced electronic devices. This, in turn, has resulted in a shortened lifespan and early obsolescence of electronic gadgets, necessitating their disposal and generating E-waste. The decrease in the longevity of electronic devices has led to a marked increase in the quantity of E-waste produced globally. The

S. Ramakrishna and B. Ramasubramanian, *Handbook of Materials Circular Economy*,
https://doi.org/10.1007/978-981-97-0589-4_5

total amount of E-waste produced in 2016 was estimated to be around 44.7 million tons, with an average of 6.1 kg per person. Additionally, it is predicted that globally, the level of e creation would rise by 3–5% year [2]. Notwithstanding the problem of quantity, e-waste poses a serious risk to the ecosystem and human health since it includes up to 1000 hazardous compounds that might have a negative impact on both. In addition to persistent organic pollutants such as aflatoxin, brominated flame retardants, polychlorinated biphenyls (PCBs), polyvinyl chloride, and fluorinated materials, toxic metals and metalloids like arsenic, cesium, lanthanum, cadmium, cerium, chromium, and copper are present in E-waste [2]. This diverse array of toxic substances has the potential to cause harmful effects on human health and the environment if appropriate management practices are not implemented (Table 5.1).

Sustainability in electronic waste (e-waste) management is about minimizing the negative impact of e-waste on the environment and human health while maximizing the value of its resources. To achieve this, e-waste should be managed throughout its life cycle, which consists of four stages: production, distribution, use, and end of life. The production stage involves designing and manufacturing electronic devices with a focus on reducing the environmental impact of production and ensuring products are designed for durability, repairability, and recyclability. Recent advances in this stage include the use of sustainable materials and eco-design practices, such as designing devices for disassembly and using renewable energy in production. The distribution stage involves the transport of electronic devices from manufacturers to retailers and consumers, with a focus on minimizing the carbon footprint of transportation and reducing packaging waste. Recent advances in this stage include the use of electric vehicles for transportation and sustainable packaging materials, such as biodegradable and compostable materials [3].

The use stage involves the operation of electronic devices by consumers, with a focus on energy efficiency, proper maintenance, and responsible disposal of devices at the end of their useful life. Recent advances in this stage include the development of energy-efficient devices, eco-labelling, and the promotion of responsible consumption behavior. The end-of-life stage involves the disposal or recycling of electronic devices, with a focus on minimizing the negative impact on the environment and human health while recovering valuable resources such as metals, plastics, and glass [4]. Recent advances in this stage include the use of innovative recycling technologies, such as hydrometallurgy and biomettallurgy, and the development of e-waste management policies and regulations. E-waste can be classified into several categories, including large household appliances, small household appliances, IT and telecommunications equipment, consumer equipment, lighting equipment, electrical and electronic tools, and toys, leisure, and sports equipment. Each category requires specific protocols and manufacturing requirements for proper disposal and recycling. To achieve sustainability in e-waste management, it is essential to implement effective policies and practices at each stage of the life cycle [5, 6]. This includes designing products for repair and recycling, promoting responsible consumption and disposal behavior, and investing in e-waste recycling infrastructure and technology. By adopting sustainable e-waste management practices, we can reduce the negative

Table 5.1 List of metals and their usage in electrical equipments and disposable health hazards

Substances	Precious metals	Component of electrical and electronic equipment	Effects on human health and environment, if not disposed after use in electronic waste
Gold	Yes	Circuit boards, connectors, microprocessors, and memory chips	Can cause soil and water pollution if not disposed of properly; Exposure can cause skin irritation, respiratory problems, and even cancer
Silver	Yes	Batteries, switches, contacts, and connectors	Can contaminate water sources and aquatic life; Prolonged exposure can lead to neurological disorders, skin irritation, and eye damage
Platinum	Yes	Hard disk drives, fiber optic cables, and LCD screens	Can release harmful gases when incinerated; Inhalation of platinum fumes can cause lung damage and asthma-like symptoms
Palladium	Yes	Hard disk drives, catalytic converters, and fuel cells	Can contaminate soil and water sources; Exposure can cause respiratory problems, skin irritation, and eye damage
Copper	No	Wires, cables, and transformers	Can cause soil and water pollution if not disposed of properly; Prolonged exposure can cause gastrointestinal problems, anemia, and liver and kidney damage
Aluminum	No	Cans, capacitors, and cables	Can contaminate water sources and aquatic life; Prolonged exposure can cause lung damage and neurological disorders
Lead	No	CRT monitors, batteries, and solder	Can contaminate soil and water sources; Exposure can cause developmental problems, anemia, and neurological disorders
Mercury	No	Fluorescent lamps, thermostats, and batteries	Can contaminate water sources and aquatic life; Exposure can cause brain damage, tremors, and kidney damage
Nickel	No	Batteries, capacitors, and connectors	Can cause skin irritation, respiratory problems, and allergic reactions; Prolonged exposure can cause lung cancer and kidney damage
Zinc	No	Batteries and coatings	Can cause soil and water pollution if not disposed of properly; Prolonged exposure can cause gastrointestinal problems and neurological disorders
Cadmium	No	Batteries, coatings, and solder	Can contaminate soil and water sources; Exposure can cause lung damage, kidney damage, and cancer
Chromium	No	Coatings, circuit boards, and hard disk drives	Can cause lung cancer, respiratory problems, and skin irritation; Exposure can also cause gastrointestinal problems and liver and kidney damage

(continued)

Table 5.1 (continued)

Substances	Precious metals	Component of electrical and electronic equipment	Effects on human health and environment, if not disposed after use in electronic waste
Vanadium	No	Batteries and coatings	Can cause respiratory problems, skin irritation, and eye damage; Prolonged exposure can cause lung cancer and neurological disorders
Cobalt	No	Batteries, magnets, and coatings	Can cause respiratory problems, skin irritation, and eye damage; Prolonged exposure can cause lung cancer and heart and kidney damage
Manganese	No	Hard disk drives and capacitors	Can contaminate soil and water sources; Prolonged exposure can cause neurological disorders and lung damage
Iron	No	Wires, transformers, and motors	Can cause gastrointestinal problems, liver damage, and heart failure; Prolonged exposure can also cause lung damage
Tungsten	No	Hard disk drives and circuit boards	Can cause lung cancer and skin irritation; Prolonged exposure can also cause neurological disorders and liver damage
Lithium	No	Batteries	Can cause respiratory problems, skin irritation, and eye damage; Prolonged exposure can also cause kidney damage and neurological disorders
Beryllium	No	Circuit boards, springs, and connectors	Can cause lung cancer and neurological disorder

impact of e-waste on the environment and human health while maximizing the value of its resources [7, 8].

When managing electronic devices, toxicity levels are essential, especially when assessing possible problems associated with their disposal. The hazardous components that are to blame for the toxicity of electronic gadgets include heavy metals and flame retardants. Toxicity levels help in determining the risks that these substances represent to the environment and to human health. The bioaccumulation factor (BAF) evaluates the concentration of a toxic chemical in an organism's environment, whereas the toxicity characteristic leaching procedure (TCLP) is used to establish if a substance is categorized as hazardous waste. These principles guide laws and rules governing the ethical disposal and recycling of electronic equipment, such as the Restriction of Hazardous Substances (RoHS) and Waste Electrical and Electronic Equipment (WEEE) Directives of the European Union. Policymakers and business leaders may work toward more sustainable and environmentally friendly methods in the manufacturing, use, and disposal of electronic gadgets by understanding the toxicity values of such devices and the parts that make them up (Table 5.2).

Electronic and electromagnetic device manufacture frequently uses plastics. They protect components from environmental harm and act as insulators for electrical

Table 5.2 TCLP. BAF and WEEF ratings of individual metals used in electronic wastes

Metal	TCLP Value (mg/L)	BAF Value (L/kg)	WEEE Value (mg/kg)
Lead	5	0.1–1.0	4
Cadmium	1.0	0.01–0.5	0.01–100
Mercury	0.2	0.001–0.1	5
Chromium	5.0	0.05–1.0	20–500
Arsenic	5.0	0.01–0.5	5
Copper	1.3	0.05–1.0	200–600
Nickel	30.0	0.1–1.0	100–300
Zinc	25.0	0.1–1.0	150–2000
Aluminum	100.0	0.1–1.0	80–500
Tin	25.0	0.1–1.0	100–200
Antimony	0.5	0.001–0.1	2–4
Beryllium	0.75	0.01–0.1	0.1–5
Cobalt	1.3	0.01–0.1	4–20
Manganese	1.0	0.1–1.0	800–900
Vanadium	7.5	0.1–1.0	30–90

wires. Just beneath the outermost plastic layer, mechanical support is frequently provided by steel wires or tapes. In limited quantities, plastics are also used in electrical products as enclosures, specialty adhesives, and protective materials (which can be either plastic or metal). One of the most widely used polymers is polyvinylchloride (PVC), which had a global output of 60 million metric tons in 2016 [9]. Nevertheless, PVC is hostile to the ecosystem, and its usage is being curtailed because of this and the dangers associated with its industrial chemicals, which are frequently brominated chemical molecules. Notwithstanding this, PVC remains a common material choice for producers due to its cheap and adaptability. Retardants are important for electrical and electronic equipment since plastics are very combustible. Retardants made of bromine are hazardous to the ecology, though. There has been a movement in latest days to use less PVC in electrical products. Electronic parts in various configurations demonstrate how widely used plastic is in the sector [10]. PVC is frequently used to insulate household electrical wire, but when it is burned to recycle copper, it releases toxic chemicals that harm the environment. Environmental impact is also caused by the PVC plasticizers known as phthalate esters. The process of pyrolysis is not frequently utilized to turn plastic trash into liquid gasoline. Due to environmental concerns, PVC is gradually being replaced with polyethylene and cross-linked polyethylene for electrical insulation. Although polycarbonate polymers are frequently used for enclosures, their use is restricted because to worries about the toxicity of bisphenol-A. In especially for Printed Circuit Boards (PCBs), PVC, and Brominated Fire Retardants (BFRs), the slow breakdown of polymers constitutes an environmental risk in landfills [11].

5.2 Classification of E-waste

A system for classifying electronic waste into six main categories was developed by the Institute of Electrical and Electronics Engineers (IEEE) and the United Nations University (UNU) called the International Technology and Engineering Educators Association (ITEEA) Classification of Electronic Waste. In addition, a standardized system called the Convention on the Control of Transboundary Movements of Hazardous Wastes and Their Disposal (CEW) codes was established by the Basel Convention to track and manage hazardous waste movement. Here are the e-waste classifications according to the ITE and CEW codes (Table 5.3).

The entire quantity of electrical and electronic equipment (EEE) used in a specific geographic area and time frame is referred to as "put-on-market." Many EEE product categories, including those for computers, televisions, refrigerators, washing machines, and more, may fall under this. There are numerous sources of information on the introduction of EEE to the market in industrialized nations, particularly in the European Union (EU), there aren't as many for emerging nations. Eurostat is one source of information on EEE sales, although emerging nations have insufficient data on the market introduction of EEE [11–13].

5.3 Recycling Strategies for Electronic Waste

The responsible management of electronic waste, or e-waste, requires the implementation of diverse techniques to dispose of electronic devices, components, and materials in an environmentally safe way. Discarded electronics like computers, phones, and televisions contain hazardous substances that may endanger human health and the environment when not adequately disposed of. Given the considerable environmental impact of the materials used in electronics, legislation has already been enacted to control their usage. It is anticipated that this tendency will continue, resulting in additional limitations on the usage of more widely used substances and additions. These restrictions may be the consequence of new rules being introduced or existing laws being strengthened, such as the RoHS (the Restriction of the use of certain Hazardous Substances in electrical and electronic equipment) Directive, which prohibits the use of certain hazardous compounds in electrical and electronic equipment [13, 14].

The Waste Electrical and Electronic Equipment Directive (WEEE) Directive lists Polybrominated biphenyls (PBBs) and polybrominated diphenylethers (PBDEs) in addition to heavy metals like lead, cadmium, arsenic, and chromium as banned materials. Except for medical equipment and monitoring and control devices, each substance has specific concentrations that cannot be exceeded in electrical or electronic items that are put on the market following July 1, 2006 [15]. The weight of a "monolithic material," which is described as a unit that cannot be physically broken down into separate materials, is used to calculate these permissible

Table 5.3 Classification of e-wastes and their corresponding ITE, CEW codes

E-waste classification	ITE code	CEW code	Examples	Additional information/ parameters
Large household appliances	LHA	Y11	Refrigerators, washing machines, air conditioners	Major appliances used for domestic purposes
Small household appliances	SHA	Y12	Toaster, coffee maker, vacuum cleaner	Small appliances used for domestic purposes
Information technology and telecommunications equipment	ITE	Y13	Computers, laptops, phones, routers	Devices used for communication and data processing, often containing valuable metals
Consumer electronics	CE	Y14	TVs, DVD players, game consoles	Electronics intended for personal or household use, often containing valuable metals and hazardous materials
Lighting equipment	LE	Y16	Fluorescent tubes, LED and CFL bulbs	Light sources used for indoor or outdoor lighting, often containing hazardous materials
Electrical and electronic tools	EET	Y45	Drills, saws, sewing machines	Tools powered by electricity or batteries used for industrial or domestic purposes
Toys, leisure, and sports equipment	TLSE	Y46	Remote-controlled toys, treadmills	Devices intended for entertainment or sports activities, often containing hazardous materials
Medical devices	MD	Y39	X-ray machines, MRI scanners	Equipment used in the medical field, often containing hazardous materials and requiring proper disposal
Monitoring and control instruments	MCI	Y48	Thermostats, smoke detectors	Devices used to measure, control, or monitor industrial or environmental processes
Automatic dispensers	AD	Y47	Vending machines, ATMs	Machines that automatically dispense products or cash, often containing valuable metals

levels. For mercury, arsenic, cadmium, hexavalent chromium, PBBs, and PBDEs, the maximum permissible concentration levels are 0.01% by weight and 0.1% by weight, respectively [15, 16].

5.3.1 Collection of E-waste

It has been common practice for many years to collect electronic garbage using traditional means, and this practice is still prevalent today. One of the most popular ways to collect e-waste is through government-sponsored recycling programs. These programs often entail set drop-off locations or scheduled pick-up services for homes and businesses, and they are administered at the regional or local scale. Another popular technique for gathering e-waste is retailer take-back programs. These programs, which include returning old devices for recycling when new products are acquired, are frequently provided by manufacturers or merchants. Local governments and non-profit organizations frequently host e-waste collection events, giving consumers a chance to get rid of their obsolete devices in one convenient location. Yet, as the volume of electronic trash increases, there is a need for e-waste collection and segregation techniques that are more effective and efficient. The usage of sensor-equipped smart bins is one such innovative tactic. These containers have the ability to monitor their level of fill and alert trash collectors when they need to be emptied. Additionally, businesses can use automated systems to classify e-waste by kind and send the items to the proper recycling facility, eliminating the need for manual sorting and processing. The use of mobile e-waste collection systems that tour different communities or businesses is another tactic. The disposal of unwanted gadgets is made simple and accessible by means of these specialized vehicles. Moreover, they may have sorting components akin to those found in smart bins, enabling the effective collection and division of e-waste [5, 11].

5.3.2 Emerging Technologies for e-waste Collection

Using drone technology to gather e-waste is a new effort that has many advantages over conventional rubbish collection techniques. Especially in rural areas or hazardous waste sites, drones with high-definition cameras and GPS tracking technologies are more effective than conventional methods for identifying and locating e-waste. Using a drone with a high-definition camera and GPS tracking technology is the method by which e-waste is collected using drone technology. The area to be scanned is flown over by the drone, and the camera records pictures of the e-waste. The drone can locate the e-waste and provide a map of the area thanks to GPS technology. To determine the type and amount of e-waste present, the drone's data is subsequently evaluated.

The project carried out by the Dubai Municipality in the UAE using drone technology for e-waste collection was implemented in 2019 [17]. The aim of the project was to reduce the environmental impact of e-waste in the region and to improve the efficiency of waste collection in remote and hazardous waste sites. The drones were equipped with high-definition cameras and GPS technology to locate and identify e-waste more efficiently. The drones were flown over the areas to be scanned, and the cameras captured images of the e-waste. The GPS technology enabled the drones to create a map of the area and identify the location of the e-waste. The data collected by the drones was then analyzed to identify the type and quantity of e-waste in the area. The success of the project can be evaluated in terms of its impact on the environment and its effectiveness in waste collection. The project was successful in reducing the environmental impact of e-waste by removing hazardous waste from remote locations and ensuring proper disposal [17]. In terms of waste collection, the use of drone technology was found to be more efficient and effective than traditional methods. The drones were able to scan larger areas in less time, enabling the collection of more e-waste in a shorter period. This increased efficiency led to a reduction in costs and improved the overall effectiveness of waste collection in the region. The cost can be high due to the initial investment in the drone and other equipment, such as cameras and GPS devices. However, the long-term benefits of using drone technology can outweigh the initial cost. Maintenance of drone technology for e-waste collection involves regular inspections and servicing of the drone and other equipment. This includes checking the battery life, cleaning the camera lens, and ensuring that the GPS device is functioning correctly. Regular maintenance can help to ensure that the drone is functioning at peak performance and minimize downtime.

5.3.3 Sorting of E-waste

Sorting and segregation techniques are essential processes in e-waste management to properly dispose of and recycle electronic waste. There are different techniques available for sorting and segregation of e-waste, including manual sorting, magnetic separation, eddy current separation, gravity separation, and optical sorting [18]. Each technique has its advantages and limitations, and the selection of a particular technique depends on the type of e-waste being processed and the end goal. Manual sorting is the most traditional and commonly used technique for e-waste segregation. It involves the manual separation of different components of e-waste by trained workers. These workers identify and separate the different materials present in e-waste, such as metals, plastics, glass, and circuit boards. This method is labor-intensive and time-consuming but effective in recovering materials that can be recycled. Initially, the electronic waste is collected from several sources such as residential areas, commercial establishments, and industrial facilities [18]. Then, the e-waste is transferred to a sorting center, where it is thoroughly inspected and classified into various categories based on the material it is composed of, such as metals, plastics, glass, and other materials. The sorting process necessitates the use of various tools, such as hammers, pliers, screwdrivers, and wire cutters, to disassemble and separate the different constituents of the e-waste [19, 20]. However, this procedure can be dangerous, as some electronic waste may contain harmful substances, like lead, mercury, and cadmium, which can pose a significant risk to both human health and the environment. Therefore, workers need to wear protective gear, such as gloves, masks, and goggles, to minimize their exposure to these harmful substances. Magnetic separation is a technique that uses a magnet to separate ferrous metals from non-ferrous metals in e-waste. The ferrous metals are attracted to the magnet, while non-ferrous metals are not. This method is useful for recovering valuable metals like iron and steel, which are abundant in e-waste. The process of magnetic separation involves placing a magnet near a mixture of materials, and as the mixture passes by the magnet, the magnetic properties of the ferrous materials cause them to be attracted to the magnet. The non-ferrous materials, which do not possess magnetic properties, are not affected by the magnet and pass by it. In the context of e-waste recycling, the mixture of materials to be separated is usually crushed and shredded to make it easier to handle. The crushed mixture is then fed onto a conveyor belt, which passes by a powerful magnet. The ferrous metals, such as iron and steel, are attracted to the magnet and are pulled out of the mixture by the magnetic force. The non-ferrous metals, such as aluminum, copper, and gold, are not affected by the magnet and continue on the conveyor belt. After the magnetic separation process, the ferrous metals are collected and sent for further processing, such as smelting or refining. The non-ferrous metals are also collected separately and sent for further processing, such as melting or refining. Eddy currents are also created by a magnetic field that induces electric currents in conductive materials, which repel them from the field. This technique is useful for separating metals which have high conductivity and are abundant in e-waste [18, 19].

The process of sorting e-waste using machine learning (ML) and artificial intelligence (AI) entails teaching a computer algorithm to identify and categorize various types of materials based on their physical and chemical qualities. By properly detecting the kind and material of electronic trash, lowering human error, and adjusting to new types of electronic waste, machine learning has the potential to improve e-waste sorting [21]. Nevertheless, putting an ML system into place necessitates a substantial financial outlay for equipment, software, and employees, in addition to massive volumes of data processing. Important variables to take into account include the limited comprehension of the system's output and the technology's limited influence on other phases of e-waste treatment. Because of this, even if ML has the potential to completely transform how e-waste is sorted, it should be a component of an all-encompassing system of e-waste management that takes into account the whole lifecycle of electronic equipment. Electronic waste is composed of a multitude of components, such as screws, cables, wires, and circuit boards, that necessitate safe and effective dismantling. Disassembly tools are integral to the e-waste recycling process by enabling the removal of these components without causing further harm [21, 22]. The industry employs various types of disassembly tools, each with unique features and technical specifications. Screwdrivers are indispensable disassembly tools employed to extract screws from electronic devices. Available in different types, such as Phillips, flathead, and Torx, with varying sizes, they correspond to the screws found in different electronic devices. Pliers, another commonly used tool, grasp and hold small components in place during removal. Available in different shapes and sizes, such as needle-nose pliers, wire-cutting pliers, and slip-joint pliers, they allow for greater precision in disassembly. Cutters are employed to sever wires, cables, and other components during the disassembly process. Various types of cutters, such as wire cutters, side cutters, and diagonal cutters, with varying sizes, match the components being cut. Heat guns are essential tools used to soften adhesives and melt solder, simplifying the extraction of components. They also aid in bending plastic components back into shape. Heat guns are available in different sizes and temperature settings, which cater to the requirements of different components. Soldering irons are necessary for melting solder and joining components together. Available in different sizes and with varying temperature settings, they match the requirements of different components. Desoldering pumps are essential tools that extract excess solder from circuit boards and components. They are designed to absorb the melted solder and gather it in a reservoir for disposal. Lastly, it is critical to employ electrostatic discharge (ESD)-safe tools when working with electronic devices. ESD can inflict damage upon electronic components; hence, utilizing ESD-safe tools is vital in preventing such damage. ESD-safe tools are made of materials that do not generate static electricity, and they are designed to discharge static electricity safely [21–23].

5.3.4 *Dismantling Components*

On a disassembly table with specialized tools and workbenches, manual e-waste dismantling can be done. For manual disassembly, it's crucial to adhere to particular structural specifications, such as employing suction hoods and collection boxes large enough to hold the disintegrated parts. Also, those engaged in manual dismantling should have the proper safety gear. Through manual disassembly, parts from TVs, screens, and PCs can be recovered. Whenever the waste components are hot and hazardous, or the procedure is laborious and risky, mechanized dismantling is required. Large-scale automated systems are used in this instance, and machines are used to break down e-waste in a regulated manner. Volatile organic compounds (VOCs), among other dangerous chemicals, may be emitted during the demolition process, which could have a negative impact on the environment and health. The researchers ran repeated sampling sessions in various locations and throughout various seasons to better understand the spatiotemporal characteristics and potential health hazards connected with VOC emissions in a typical e-waste dismantling area. It was stated in previous studies that compounds with cancer risks above 1.0×10 – 4 were classified as "definite risk," those between 1.0×10^{-5} and 1.0×10^{-4} as "probable risk," those between 1.0×10^{-5} and 1.0×10^{-6} as "possible risk," and those below 1.0×10^{-6} as "negligible." It was observed that outside the EP, 1,2-DCA, 1,2-DCP, and 1,3-butadiene had cancer risks above 1.0×10^{-6}, and benzene

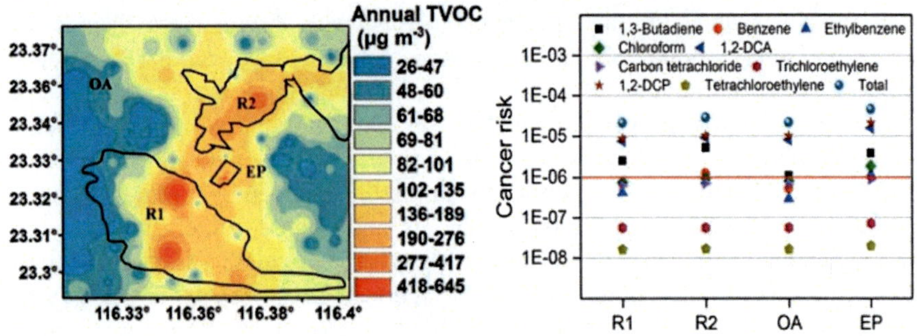

Fig. 5.1 Hazardous gas emission from e-waste dismantling [24]

slightly exceeded this value in R2, primarily due to long-range transport and vehicular emissions, as shown in Fig. 5.1. Conversely, inside the EP, it was found that ethylbenzene and chloroform posed a definite cancer risk, indicating a higher risk associated with e-waste dismantling activities [24].

The recycling of electronic waste has led to the production of non-biodegradable and heavy metal garbage in emerging nations like China, India, Pakistan, and Ghana. The increase in the creation of e-waste on a global scale has made this problem worse. The soil was discovered to contain different levels of pollution in the unofficial e-waste disposal places, mostly at low and medium levels, which are treatable with bioremediation techniques.

A study examined the impact of e-waste dismantling with eco-friendly technologies in a South China industrial park (established in 2015) on the surrounding area and human health risks. Soil analysis revealed higher concentrations of flame retardants, notably PBDEs (specifically BDE209) and OPEs (particularly triphenyl phosphate), in the industrial park compared to the surrounding area. DP's fanti value remained stable at 0.75. The study highlights the need for future attention to PBDEs and OPEs as primary contributors to the hazard quotients for children in the park, Fig. 5.2 [25].

5.3.5 Advanced Recycling Techniques

Biometallurgical recycling is a revolutionary technique that is increasingly gaining traction in the e-waste recycling industry. It involves the utilization of microorganisms to extract valuable metals from electronic waste. The process commences with the collection and sorting of e-waste, which is then shredded to obtain a fine powder. The powder is then mixed with water and inoculated with bacteria or fungi that can dissolve the metals present in the e-waste [26]. These microorganisms excrete organic acids that dissolve the metals, which can then be extracted from the solution. The remaining waste can be disposed of safely without posing any environmental threat. This technique has several advantages over traditional recycling methods, as it requires less energy and does not produce harmful emissions. Additionally, it

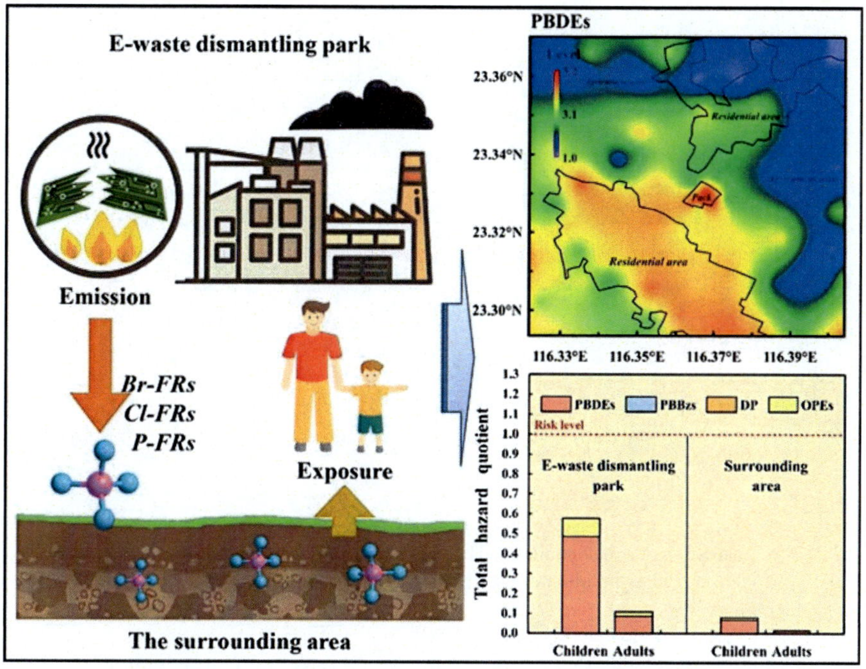

Fig. 5.2 Flame retardants in e-waste park soils: Distributions, sources, and health risks [25] (*PBD* polybrominated diphenyl ethers, *PBB* polybromobenzenes, *DP* Dechlorane plus, *OPEs* organophosphate esters)

has the potential to extract a wider range of metals from e-waste, including rare earth elements that are crucial to produce high-tech devices. Bioleaching is a similar process of using microorganisms to extract metals from ores and other materials, while biomettallurgy is a more specialized form of bioleaching that is used specifically for the extraction of metals from e-waste. Researchers at the University of Edinburgh are currently studying the effectiveness of biometallurgical recycling for e-waste and have reported promising results in laboratory tests. However, more research is necessary to scale up the process for industrial applications [27]. Figure 5.3 illustrates bioremediation techniques for typical e-waste pollutants, aiding stakeholders in adopting suitable technologies.

The microbial cell suspension used in the one-step bioleaching procedure is taken from the exponential growth stage and added to a bioleaching medium that is appropriate for the particular type of e-waste being processed. In this procedure, ferrous iron (Fe^{2+}) is oxidized to ferric iron (Fe^{3+}) and protons, which dissolve any imbedded metals in the e-waste (Fig. 5.3). Nevertheless, the one-step bioleaching procedure can only be used at low pulp density, typically between 1 and 10% (w/v), because the presence of harmful chemicals in e-waste can hinder microbial growth. Direct growing of microbes in the presence of e-waste is therefore not advised because the

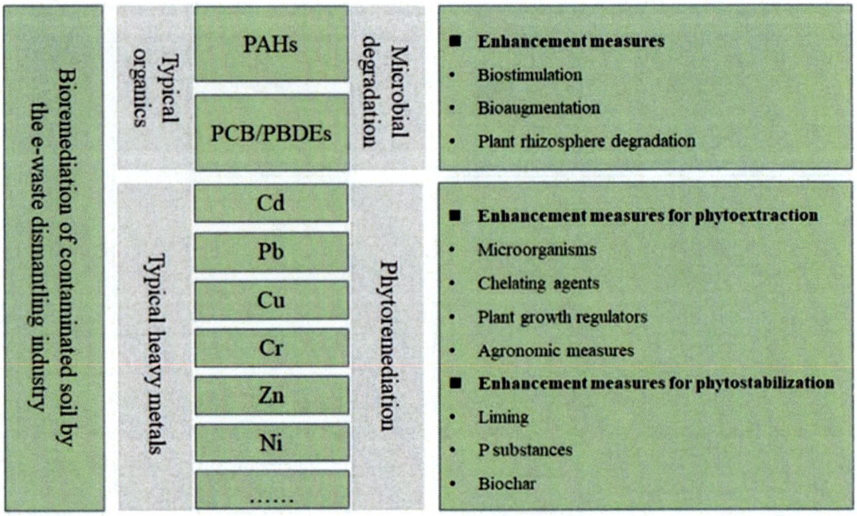

Fig. 5.3 Bioremediation of soil contaminated by pollutants, particularly organics and heavy metals from e-waste, is discussed using literature metrics and remediation strategies [26]

hazardous materials in e-waste can inhibit microbial development, which lowers the efficiency of metal extraction (Table 5.4).

It has been founded that chemolithotrophs use specific enzymes to speed up the oxidation of inorganic substances, including hydrogenases and sulfur oxidases. A few chemolithotrophs can also adopt a method known as lithotrophic carbon fixation, in which they use carbon dioxide as their only carbon source, according to recent studies. Similar to this, it has been discovered that a range of enzymes and metabolic pathways are used by organotrophs to break down chemical molecules. For instance, many species use the citric acid cycle and glycolysis to break down glucose, which is a crucial process for producing ATP. Moreover, recent research has demonstrated that some organotrophs could use other electron acceptors in the absence of oxygen, such as nitrate or sulfate.

Supercritical fluid (SCF) extraction is an advanced method of e-waste recycling that employs supercritical fluids, such as carbon dioxide, as a solvent to recover valuable materials. A supercritical fluid is a substance that is maintained at a temperature and pressure above its critical point, exhibiting properties of both a gas and a liquid. The extraction process employs a closed-loop system, wherein the e-waste is placed in a vessel and the supercritical fluid is circulated through it. The supercritical fluid is capable of dissolving both organic and inorganic materials, including metals, from the e-waste. After extraction, the pressure is released, and the supercritical fluid evaporates, leaving behind the extracted materials. Supercritical fluid extraction is especially effective in extracting metals, such as copper, gold, and silver, from e-waste. Compared to conventional extraction techniques, this method offers numerous benefits, including reduced environmental impact, enhanced efficiency,

Table 5.4 Bio-based recycling source and reaction mechanism for leaching individual metals

Metal	Bacteria/fungi	Energy source	Reaction equation
Aluminum (Al)	Acidithiobacillus ferrooxidans, Acidithiobacillus thiooxidans, Aspergillus niger	Chemolithotroph, Organotroph	$4Al + 12H_2SO_4 + 3O_2 \rightarrow 4Al_2(SO_4)_3 + 6H_2O$
Copper (Cu)	Acidithiobacillus ferrooxidans, Acidithiobacillus thiooxidans, Leptospirillum ferriphilum	Chemolithotroph, Organotroph	$CuFeS_2 + 2O_2 + 2H_2SO_4 \rightarrow CuSO_4 + FeSO_4 + 2H_2O + 2SO_2$
Gold (Au)	Acidithiobacillus ferrooxidans, Acidithiobacillus thiooxidans, Sulfolobus metallicus	Chemolithotroph	$4Au + 8CN- + O_2 + 2H_2O \rightarrow 4[Au(CN)2]- + 4OH-$
Iron (Fe)	Acidithiobacillus ferrooxidans, Acidithiobacillus thiooxidans, Leptospirillum ferrooxidans	Chemolithotroph, Organotroph	$4FeS_2 + 15O_2 + 14H_2O \rightarrow 4Fe(OH)3 + 8H_2SO_4$
Lead (Pb)	Desulfotomaculum nigrificans, Bacillus sphaericus	Chemolithotroph, Organotroph	$PbS + 2O_2 + 2H_2O \rightarrow PbSO_4 + 2H_2O_2$
Nickel (Ni)	Acidithiobacillus ferrooxidans, Acidithiobacillus thiooxidans, Leptospirillum ferriphilum, Aspergillus niger	Chemolithotroph, Organotroph	$NiS + 2O_2 + 2H_2O \rightarrow NiSO_4 + 2H_2SO_4$
Silver (Ag)	Acidithiobacillus ferrooxidans, Acidithiobacillus thiooxidans, Pseudomonas stutzeri	Chemolithotroph, Organotroph	$2Ag_2S + 8CN- + O_2 + 2H_2O \rightarrow 4[Ag(CN)2]- + 2SO_4- + 4OH-$
Tin (Sn)	Aspergillus niger, Penicillium simplicissimum	Organotroph	$SnO_2 + 4C \rightarrow Sn + 2CO_2$
Zinc (Zn)	Acidithiobacillus ferrooxidans, Acidithiobacillus thiooxidans, Leptospirillum ferriphilum	Chemolithotroph, Organotroph	$ZnS + 2O_2 + 2H_2O \rightarrow ZnSO_4 + 2H_2SO_4$
Palladium (Pd)	Acidithiobacillus ferrooxidans, Acidithiobacillus thiooxidans, Pseudomonas aeruginosa	Chemolithotroph, Organotroph	$PdS + 4Fe^{3+} \rightarrow Pd^{2+} + 4Fe^{2+} + S$

and higher yields. Moreover, the extracted metals possess higher purity and can be directly reused in the production of new electronic devices, reducing the reliance on virgin materials.

Due to its unique physical and chemical characteristics, supercritical water is an efficient catalyst for organic reactions and can offer a fresh method of waste treatment. It exhibits behavior akin to organic solvents and provides a lot of hydrogen and hydroxyl ions. Supercritical CO_2 is a more beneficial solvent than supercritical water, which has a comparatively high supercritical point due to water's high mass transfer efficiency, perfect miscibility with gaseous reactants, and simplicity in product separation. A new, environmentally acceptable method for studying chemical reactivity is to use CO_2 as a solvent. The kinetics of SCF oxidation of organic compounds is generally believed to follow a first-order or pseudo-first order model relative to the concentration of organic materials, with oxidation rates being independent or weakly dependent on oxidant concentration. E-waste treatment can be achieved through two reactor designs, namely, tank and tubular reactors. Tubular reactors are commonly used for the treatment of liquid and sludge, providing high treatment capacity and ease of operation. Tank reactors, including sequencing batch and semi-continuous reactors, are typically utilized for the oxidation and extraction of solid-state raw materials. While tubular reactors are popular for their simplicity, they are not suitable for treating large solids, Fig. 5.4 [28].

Hybrid recycling techniques refer to the combination of two or more recycling methods to create a more effective and efficient process. These techniques are becoming increasingly popular in the e-waste recycling industry as they can provide better recovery rates and help to reduce the environmental impact of the recycling process.

Umicore is a Belgian recycling firm that specializes in the recovery of valuable metals from e-waste. To achieve this, the company uses a combination of hydrometallurgy and pyrometallurgy [29]. Hydrometallurgy involves using a liquid solvent to extract metals from ores, concentrates, or other materials, while pyrometallurgy involves high-temperature processes like smelting or roasting. By combining these two methods, Umicore can recover metals that would be difficult to extract using a single technique. EnviroLeach Technologies is a Canadian company that uses a

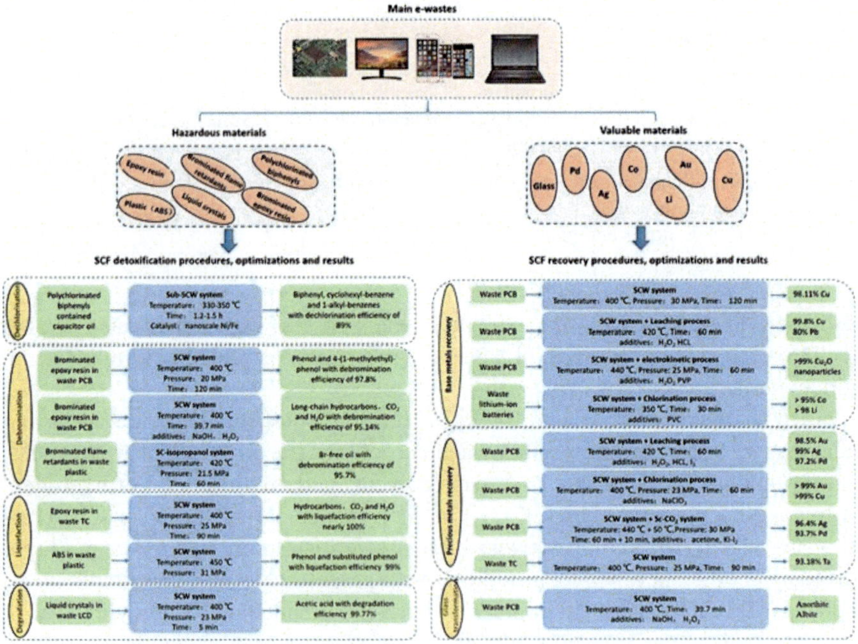

Fig. 5.4 Supercritical fluid (SCF) detoxification and recovery procedures for recycling of e-waste [28]

combination of chemical leaching and electrochemical processing to extract metals from e-waste. TES-AMM (Sustainable recycling) in Singapore that is focused on sustainable electronics recycling. The company uses a combination of three different techniques to extract valuable materials from e-waste. The first technique, mechanical processing, involves breaking down e-waste into smaller pieces, typically by shredding or grinding it. This process can separate components based on their size or density, allowing for further processing to recover valuable materials. The second technique, chemical leaching, involves the use of chemicals to dissolve the valuable materials in e-waste. The resulting solution can be treated to recover the dissolved metals, leaving behind the non-valuable components. Finally, the third technique, pyrometallurgy, involves the use of high temperatures to melt down e-waste and separate out the valuable materials. This process is often used for materials that cannot be easily extracted through other methods, such as certain metals and alloys. TES-AMM was founded in 2005 and has since grown to become a leading provider of sustainable e-waste management solutions in the Asia–Pacific region. E-waste collection, transportation, and processing are among the services provided by TES-AMM. The business offers services for securely destroying data and disposing of IT assets, making sure that sensitive information is handled safely and sustainably. Gold, silver, and palladium are among the precious metals that TES-AMM recycles and refines from e-waste. TES-AMM offers counselling and auditing services in addition to its core recycling services to assist businesses and organizations in managing their e-waste sustainably. The business places a high priority on sustainability, and through fostering the circular economy, cutting waste, and using less hazardous materials, it hopes to lessen the impact of e-waste on the environment.

5.4 Alternate Materials and Solutions

One of the most pressing concerns with electronics is the use of rare earth metals, which are essential components in the production of electronics. These metals are in short supply and difficult to extract, often leading to environmental degradation and human rights abuses in the countries where they are mined. Mycelium, the vegetative part of a mushroom, can be used to create biodegradable and sustainable packaging for electronic devices. Founded in 2007, Ecovative Design is based in Green Island, New York. The company produces mycelium-based materials that can be used for a range of applications, including packaging, insulation, and consumer products.

Given the advanced state of roll-to-roll printing methods, paper is a highly suitable substrate for printed electronics. The porous structure of the paper fiber network results in high surface roughness, which can be advantageous for energy storage devices requiring large surface area for easy absorption of electrolyte and binding with nanomaterials. When conductive materials such as carbon nanotubes, silver nanowires, metal oxides, graphene, and conductive polymers are incorporated within the paper, it achieves sufficient electrical conductivity for the subsequent hybrid materials used as substrates in devices like supercapacitors. Interestingly, some devices prepared on such modified conducting paper substrates have performed comparably with analogous devices deposited on flat conducting polymer substrates. However, the high roughness and porosity of paper can prove detrimental to thin film devices, especially those utilizing organic materials as semiconducting or dielectric layers. This problem can be solved by coating or laminating with wax, kaolin, starch, latex, and various polymers such as polyethylene, polypropylene, polyurethane, and PVA. Thus, paper substrates have been extensively used in various electronic devices including solar cells, organic LEDs, biobatteries, sensors, transistors, displays, and radio frequency identification devices. Thin wood slices cut transverse to the plane of development can be used to create transparent paper substrates using a simple technique devised by Zhu et al. [30]. The ensuing tubular formations, known as lumens, were aligned parallel to the path of growth when the lignin was removed, and when they were squeezed along the lumen axis, they randomly collapsed. The resultant material featured a thick layered structure, an isotropic network of fiber strands, and was resistant to optical diffraction. The absorbance of this "uniaxial sheet" was almost 90%. Ohmic contact was shown using a straightforward device made by connecting a graphene flake with two gold electrodes placed through a shadow mask. The instrument displayed a bipolar transistor function in response to the addition of a gate electrodes and an electrolytes [31].

As a substrate for OLED displays, Bae et al. created a paper based on chitin that demonstrated good transmission in the visible spectrum. Silk fibers are being investigated for potential uses in biomedical electronics because they are also biocompatible and biodegradable. Silk contains a protein called fibroin, which can be used to create paper-like green technology substrates with great bioactivity and gradual disintegration. Additional natural substances with the potential to be used as biologically compatible and biodegradable substrates include shellac, hard gelatin, collagen, chitosan, alginate, and dextran. The ecological stability and suitability of certain polymers for several methods of device manufacture are, however, constrained. As a result, attention is being drawn to artificial substrates with adjustable qualities. In the realm of green electronics, iron, magnesium, zinc, and their alloys and oxides are commonly employed as electrodes and contacts due to their biocompatibility and ease of disposal. Conducting polymers like polypyrrole, polyaniline, and polythiophene can also serve as electrode materials after doping to attain higher conductivity. Carbon nanotubes offer a unique option due to their special properties and biocompatibility. Nonetheless, the use of organic materials as electrodes is restricted by challenges in fabrication and concerns about toxicity. Overall, metals are still the preferred choice for electrode materials in green electronics. To degrade conducting polymers, one approach is to create a composite that comprises conducting polymers and a biodegradable insulating matrix.

5.4.1 Shared Economy Model

The shared economy circular model is a business strategy that aims to establish a more sustainable and circular economy by promoting the sharing of resources, reducing waste, and encouraging the reuse and recycling of goods. It is grounded on the principle that a product's life cycle should be extended by reusing and recycling instead of disposing of it after use. It is also referred to as the "circular sharing economy" or the "circular economy sharing model." To achieve this model, various companies and programs work towards facilitating the sharing of resources such as vehicles, housing, tools, and equipment among individuals and businesses. This sharing can occur in various ways, including ride-sharing, home-sharing, tool-sharing, and co-working spaces. The goal is to maximize the utilization of existing resources, minimize waste, and encourage a more sustainable way of living. Companies and programs that operate within the shared economy circular model include renowned platforms such as Airbnb, Uber, Lyft, and Zipcar, which enable individuals to share their homes and cars for short-term rentals or rides. Additionally, WeWork provides co-working spaces that allow individuals and businesses to share office space, reducing the need for separate and individual office spaces. Apart from these notable corporations, smaller, local businesses and programs are emerging within the shared economy circular model. For instance, some cities have bike-sharing programs, allowing people to rent bicycles for a brief period, reducing the necessity for people to own bikes. Likewise, there are tool-sharing programs that enable people to borrow tools for home repair and maintenance projects, reducing the need for individuals to purchase and store their tools.

Grover is a German-based start-up that operates in the shared economy of electronics by providing a subscription-based model for renting tech products to individuals and businesses. Grover offers a vast range of electronics, including smartphones, laptops, gaming consoles, cameras, and other accessories. The Grover model allows customers to subscribe to a monthly plan, which gives them access to a range of electronics products that they can rent for a certain period. Customers can choose the duration of the rental period, and they have the option to extend the rental period or purchase the product outright at any time. Grover's pricing structure is based on the rental duration, with longer rental periods costing less per month. To manage its inventory and track the usage of its products, Grover uses a combination of IoT and cloud-based technologies. Each device that Grover rents out is equipped with sensors that collect data on usage patterns, such as the number of hours the device is used and the type of applications that are run. This data is then transmitted to Grover's cloud-based platform, where it is analyzed to optimize inventory management and pricing. Grover uses statistical models and machine learning algorithms to analyze the usage data and predict the demand for its products. This allows the company to optimize its inventory levels and ensure that it has the right products available for its customers. Grover also uses predictive analytics to identify potential issues with its products before they occur, which allows the company to proactively address any problems and minimize downtime.

5.4.2 Products-as-a-Service (PaaS) Model

Once upon a time, a new business model emerged in the world of electronics: the Products-as-a-Service (PaaS) model. This model offered customers the opportunity to access electronic products and services on a subscription basis, instead of buying them outright. The PaaS model was designed to promote the circular economy, reduce waste, and provide customers with more flexible and affordable options. The PaaS model works by allowing customers to pay a recurring fee for access to a range of electronic products and services. This includes everything from smartphones and laptops to home automation systems and renewable energy solutions. The products and services are typically provided by a third-party provider who handles maintenance, repairs, and upgrades as needed. The PaaS model includes customizable packages that allow customers to choose from a range of products and services to create a package that meets their specific needs. The provider is responsible for maintaining and repairing the products and services as needed, and customers may have the option to upgrade to newer or more advanced products and services over time. One company that has adopted the PaaS model for electronics is Philips, a well-known brand in the world of lighting solutions. Philips Pay-per-Lux lighting service offers businesses and public organizations access to high-quality LED lighting solutions on a subscription basis. The service includes installation, maintenance, and replacement of the lighting fixtures, as well as ongoing support and energy management services. According to a Philips case study, the Pay-per-Lux service has helped customers reduce their energy consumption by up to 75% and their carbon emissions by up to 57%. The service has also provided customers with greater flexibility and control over their lighting solutions, as well as cost savings and reduced maintenance requirements.

5.4.3 Product Ownership Model

The product ownership model is a sustainable approach to product ownership that integrates the principles of Product Life Extension (PLE) and Design for Recyclability (DFR). PLE involves designing and manufacturing products with a longer lifespan by incorporating repair and upgrade strategies. DFR, on the other hand, focuses on designing products with the end-of-life stage in mind, making them easier to disassemble and recycle. This model aims to create a circular economy where products are produced, used, and disposed of sustainably and responsibly. The process starts with designing products that can be easily repaired, upgraded, and disassembled at the end of their lifespan using durable and high-quality materials that can withstand wear and tear. During the production phase, sustainable production methods and materials are used, and waste is minimized. In the usage phase, manufacturers provide repair and upgrade services to extend the product's lifespan. At the end-of-life stage, the product is disassembled and recycled, with the materials being repurposed into new products. Companies such as Fairphone and IKEA have adopted this model. Fairphone designs its ethical and sustainable smartphones to be easily repaired and upgraded and provides repair services to extend the lifespan of its products. IKEA has committed to using only renewable and recycled materials in its products by 2030 and has implemented a circular business model where customers can return their used furniture to be resold or recycled.

5.5 Organic Electronics

Due to its interlinked design, which alternates single and double bonds along the polymer's mainchain, conductive polymers have demonstrated success in organic electronics. The initial materials utilized were conductive synthetic polymers like PPy, PPV, PANI, PEDOT, and PT. With time, a wide range of additional materials with top-notch conjugation structures have entered the category of organic semiconductors. In recent decades, organic semiconductors have achieved tremendous advancements in printed, flexible, and organic electronics. Several of the -conjugated motifs that have been employed as building blocks in organic semiconductors, such as DPP, isoindigo, NDI, PDI, BT, benzodithiophene, carbazole, and TPA, have developed into different fields of study. Organic materials that can self-heal when damaged are a promising development in the field of materials science, with potential applications in various industries, including electronics, aerospace, and automotive. These materials are designed to repair themselves automatically when they sustain damage, without the need for external intervention. The tubular fibers that make up the majority of the mesoporous substance known as balsa wood are aligned along the direction of the tree's growth. The fibers have a polygonal cross-section and a cell wall thickness of roughly 1.8 m. Their diameter ranges from 20 to 45 m. Natural wood has a light-yellow hue because of lignin, a phenolic biopolymer that is present in every cell wall but is concentrated in the center lamella of the compound and the corners of the cell wall. Balsa wood's honeycomb structure is maintained through chemical processing, but its colour is eliminated, and its lignin concentration is reduced from 24.5 weight percent to 1.6 weight percent. Moreover, the treatment eliminates 18.2% of the cellulose content and half of the hemicellulose.

5.5.1 Organic Field Effect Transistors (OFET)

OFETs are complex electronic devices with three terminals, including an organic semiconductor layer, a gate electrode, a gate dielectric layer, and S/D electrodes. These terminals need a substrate to be supported. The relationship between the position of the gate and S/D electrodes in relation to the semiconductor layer is what allows OFETs to be divided into four different varieties, including Bottom-gate: Top and bottom contact, Top-gate: Top and bottom contact. Moreover, unique arrangements like side-gates and dual-gates have been noted. OFETs consist of a substrate, a dielectric layer, and electrodes. Each of these components plays a crucial role in the functioning of the transistor. The substrate is typically made of a rigid material such as silicon, glass, or quartz. The substrate needs to be smooth and flat to provide a good surface for the deposition of the other layers. Additionally, the substrate should have good thermal and mechanical stability to prevent deformation during the fabrication process. The choice of substrate material depends on the desired application and the manufacturing process. The dielectric layer is sandwiched between the two electrodes and serves as a gate insulator. It is made of materials with high capacitance, low leakage current, and high stability. The dielectric material should have a high dielectric constant to allow for efficient charge transfer from the gate electrode to the semiconductor layer. Some of the commonly used dielectric materials in OFETs include silicon dioxide (SiO_2), aluminum oxide (Al_2O_3), and polymeric materials such as poly(methyl methacrylate) (PMMA) and polyimide (PI). The electrodes in OFETs are used to apply a voltage to the gate and drain, and to collect the current flowing through the device. The choice of electrode material depends on the application, the manufacturing process, and the desired device performance. Some of the commonly used electrode materials in OFETs include gold (Au), silver (Ag), aluminum (Al), and copper (Cu). These materials are chosen for their high conductivity, low resistance, and compatibility with the other layers in the device. In addition to the substrate, dielectric, and electrode materials, OFETs can also contain a semiconductor layer. The semiconductor material is responsible for conducting the charge carriers from the source to the drain when a voltage is applied to the gate. Some commonly used semiconductor materials in OFETs include pentacene, tetracene, and polymeric materials such as poly(3-hexylthiophene) (P3HT) and poly(3-alkylthiophene) (P3AT).

The top-contact (TC) configuration of OFETs entails growing or depositing the organic semiconductors directly onto the dielectric layer, followed by the depositing of the S/D electrodes. This procedure guarantees the formation of high-quality crystal lattices and a uniform contact interface between the semiconductor and dielectric materials. However, because the organic semiconductors are grown or deposited on the S/D electrodes and gate dielectrics in the bottom-contact (BC) configuration, there are structural and performance differences. In contrast to the BC arrangement, the TC configuration demonstrates improved characteristics such a greater contact area and lower contact resistance between the semiconductors and electrodes. Due to the restricted mask technologies available, the TC configuration is difficult to implement practically and is therefore unsuitable for large production. A perpendicular electric field is used by OFETs to control the charge density in their active channels. It is necessary to extract information from the transfer (ISD-VG) and output (ISD-VSD) curves in order to better understand the field-effect characteristics of OFETs. These parameters include the threshold voltage (Vth), charge mobility (), Ion/Ioff, and subthreshold slope (SS). It is important to carefully evaluate variables for Flexible OFETs (FOFETs), such as device thickness, curvature radius, stretching strain, and electrical stability during mechanical deformations. Rubrene, C8-BTBT, and 2,9-didecyl-dinaphtho[2,3-b:20,30-f]thieno[3,2-b]thiophene (C10-DNTT) are examples of organic compounds having mobility values more than 10 cm^2 V1 s1 that are comparable to polycrystalline silicon FETs. Due to their negligible grain boundaries and imperfections, organic single crystals in particular are very interesting for exploring the link between electrical properties and mechanical performance. For examples, Reyes-Martinez examined the effects of physiological deformations on carrier mobility using rubrene as a benchmark semiconductor. They discovered that structural compression has the opposite impact of tension, having the reverse consequence of increasing intermolecular distance and decreasing carrier mobility. Carrier mobility was also impacted by the net strain at the dielectric/semiconductor contact. Although there are fewer examples of organic semiconductors than polymer semiconductors, research on organic compounds for making FOEFTs has drawn a lot of attention. Rubrene, C8-BTBT, and 2,9-didecyl-dinaphtho[2,3-b:20,30-f]thieno[3,2-b]thiophene (C10-DNTT) are examples of organic compounds having mobility values more than 10 cm^2 V1 s1 that are comparable to polycrystalline silicon FETs. Due to their negligible grain boundaries and imperfections, organic single crystals in particular are interesting for exploring the link between electrical properties and mechanical performance. For examples, Reyes-Martinez examined the effects of physiological deformations on carrier mobility using rubrene as a benchmark semiconductor. They discovered that structural compression has the opposite impact of tension, having the reverse consequence of increasing intermolecular distance and decreasing carrier mobility. Carrier mobility was also impacted by the net strain at the dielectric/semiconductor contact. Although there are fewer examples of organic semiconductors than polymer semiconductors, research on organic compounds for making FOEFTs has drawn a lot of attention.

Polymer semiconductors have advantages over organic small molecules such as higher molecular weight, controlled molecular structure, mechanical flexibility, and film-assembling capability. However, their highly ordered lamellar structure and high crystallinity can decrease their mechanical stretchability. P3HT and pBTTT are two polymer semiconductors with different molecular packing geometries, where pBTTT has higher carrier mobility but lower crack onset strain compared to P3HT. Increasing the flexibility and stretchability of polymer semiconductors is challenging, but strategies include designing multiblock copolymers, incorporating alkyl side chains, regulating molecular weight and regioregularity, introducing conjugated carbon cyclic nanorings or blending with elastomers. Geometrically structuring brittle polymer semiconductors through strain-engineering designs is also a viable approach. Blending processes require careful attention to surface energies, molecular weight, blending ratio, solvents, and processing methods.

5.5.2 Organic Photovoltaics

Organic photovoltaics (OPVs) are a type of solar cell that convert sunlight into electricity using organic materials. They offer several advantages over traditional silicon-based solar cells, such as flexibility, lightweight, and low-cost manufacturing.

There are two main types of OPVs: single-junction and tandem-junction. Single-junction OPVs use a single layer of a polymer or small molecule as the active material, while tandem-junction OPVs use two or more layers of different materials to absorb a broader range of the solar spectrum. Recent advancements in materials science have led to the development of new organic materials with improved efficiency and stability. For example, non-fullerene acceptors, such as perylene diimides and IDIC derivatives, have been found to increase the power conversion efficiency (PCE) of OPVs to over 18%. OPVs consist of several parts, including a transparent conductive electrode, an active layer, an electron transport layer, and a metal electrode. When sunlight hits the active layer, it generates excitons, which are electron–hole pairs. The excitons are then separated into free charges, and the electrons are collected at the metal electrode while the holes are collected at the transparent conductive electrode. The mechanism of working of OPVs is based on the principles of photovoltaic effect and charge transport. The active layer of OPVs absorbs photons from sunlight and converts them into excitons, which are then separated into free charges. The electron and hole transport layers facilitate the transport of these charges to the electrodes, where they are collected and used to generate electricity.

One commonly used material for hole transport layers (HTLs) is poly(3,4-ethylenedioxythiophene): polystyrene sulfonate (PEDOT:PSS). However, its acidity and hygroscopic nature can degrade the device performance over time. To overcome these limitations, other materials have been explored such as metal oxides like molybdenum trioxide (MoO_3) and vanadium pentoxide (V_2O_5), as well as conductive polymers such as poly(3-hexylthiophene) (P3HT). For electron transport layers (ETLs), fullerene derivatives such as [6, 6]-phenyl-C61-butyric acid methyl ester (PCBM) have been widely used due to their high electron mobility and good compatibility with the active layer. However, fullerene-based ETLs have several drawbacks, including limited light absorption and poor stability. To address these issues, alternative materials have been explored such as non-fullerene acceptors (NFAs) like ITIC and Y6, as well as metal oxides like zinc oxide (ZnO) and titanium dioxide (TiO_2). In addition to the aforementioned materials, other recent interests in HTL and ETL materials include conducting polymers such as poly(9,9-dioctylfluorene-co-bithiophene) (F8T2) and small molecules like 2,9-dimethyl-4,7-diphenyl-1,10-phenanthroline (BCP). These materials have shown promising results in improving device performance and stability.

5.5.3 *Organic Memory Devices and Organic LEDs*

Organic memory devices (OMDs) are electronic devices that use organic materials to store and retrieve information. OMDs are similar to conventional memory devices, such as flash memory, but use organic materials instead of traditional inorganic materials. OMDs consist of a substrate, a bottom electrode, an organic layer, a top electrode, and a dielectric layer. The organic layer is typically a thin film of a conjugated polymer or small molecule material that is capable of storing charge. When a voltage is applied across the electrodes, charge is injected into the organic layer and stored as trapped charges in localized states. The trapped charges can be read out by applying a voltage to the electrodes and measuring the resulting current. OMEDs work based on the switching of resistance or capacitance in the organic layer between two electrodes. This resistance or capacitance switch can be used to represent digital information, with a high resistance or capacitance state representing a binary 0 and a low resistance or capacitance state representing a binary 1. OMEDs are typically fabricated using solution-based techniques such as spin-coating or inkjet printing. The active layer of the device is made of an organic material that can switch between high and low resistance or capacitance states. The most commonly used organic materials for OMEDs are polymers and small molecules such as organic semiconductors. To improve the performance of OMEDs, several strategies have been employed. One such strategy is to modify the interface between the organic layer and the electrodes using self-assembled monolayers (SAMs) or interfacial layers. These modifications can improve charge injection, reduce contact resistance, and enhance the stability of the device. Another approach to improving OMED performance is to use hybrid materials. In hybrid OMEDs, the organic material is combined with inorganic materials such as metal oxides or nanoparticles. These hybrid materials can provide better charge transport properties, enhanced stability, and improved memory performance. In addition, researchers are exploring new types of organic memory devices, such as resistive switching devices and phase change memory devices.

Organic Light Emitting Diodes (OLEDs) are devices that emit light when an electric current passes through them. These devices are made up of organic materials, which are carbon-based compounds, and can be used to create thin, flexible, and energy-efficient displays. The basic mechanism of an OLED involves the flow of electrons from the cathode to the anode when an electric current is applied. The electrons combine with holes in the organic semiconducting layer to form excitons, which migrate to the interface between the organic semiconducting layer and the emitting layer. When the excitons decay, they release energy in the form of light. Several concepts are important in OLED technology, including electroluminescence, heterojunction, and hole and electron injection. Electroluminescence is the process by which light is produced in an OLED when an electric current is passed through it. Heterojunction refers to the interface between two materials with different electronic properties, which in an OLED is where light is produced. Hole and electron injection are the processes by which electrons and holes are introduced into the organic semiconducting layer of an OLED, typically by applying a voltage across the device. The basic parts of an OLED include the anode, cathode, organic layers, substrate, and encapsulation layer. The anode is the positive electrode, typically made of a transparent conductive oxide, while the cathode is the negative electrode, usually made of a metal such as aluminum or calcium. The organic layers consist of organic semiconductors and conducting layers, while the substrate is the material on which the OLED is deposited. Finally, the encapsulation layer is a protective layer that prevents moisture and oxygen from entering the OLED and degrading its performance.

5.6 IT Enabled Electronics

IT-enabled electronics have become an integral part of our daily lives, enabling us to connect, communicate, and perform a wide range of tasks with ease. These electronics are made of a wide range of materials, including semiconductors, conductive materials, and dielectric materials. Semiconductors are materials that have an electrical conductivity between that of a conductor and an insulator, and are typically made of silicon, germanium, or gallium arsenide. Conductive materials allow electricity to flow through them and are typically made of metals such as copper, silver, or gold. Dielectric materials do not conduct electricity but are used to insulate components from each other, and are typically made of materials such as silicon dioxide or aluminum oxide. The basic mechanism of IT-enabled electronics involves the manipulation of electrical signals to perform specific tasks. This is achieved through a combination of hardware and software components, including microprocessors, memory, and input/output devices. Microprocessors are the "brains" of the device and are responsible for processing data and executing instructions. Memory is the component that stores data and instructions, while input/output devices allow the user to interact with the system, either by providing input (such as a keyboard or mouse) or output (such as a display or speaker).

Several important concepts are involved in IT-enabled electronics, including digital electronics, integrated circuits, and Moore's Law. Digital electronics refer to the use of binary digits (0s and 1s) to represent data and instructions, and are the basis of modern computing. Integrated circuits are miniaturized electronic circuits that are made up of multiple components (such as transistors, resistors, and capacitors) on a single chip. Moore's Law is the observation that the number of transistors on a chip (and hence the processing power of computers) doubles roughly every 2 years. Advancements in IT-enabled electronics have been numerous and have enabled the development of wearable technology, the Internet of Things (IoT), and artificial intelligence (AI). Wearable technology includes devices such as smartwatches and fitness trackers, which are designed to be worn on the body and provide real-time feedback on a user's health and fitness. The IoT refers to the interconnection of everyday devices (such as refrigerators, thermostats, and light bulbs) via the internet, enabling communication between them and with cloud-based services to provide a wide range of functionality. AI is the use of algorithms and machine learning techniques to enable computers to perform tasks that would normally require human intelligence, such as image recognition and natural language processing. Examples of IT-enabled electronics include smartphones, which combine the functionality of a computer, camera, and communication device into a single handheld device. Smart homes are equipped with IoT devices, such as smart thermostats, security cameras, and voice assistants. Autonomous vehicles use a combination of sensors, cameras, and AI algorithms to navigate roads and highways without human intervention. These examples demonstrate the significant impact that IT-enabled electronics have had on our lives and the ongoing advancements that will continue to shape our future.

5.7 Global Initiatives and Policies

The United Nations Environment Programme's (UNEP) Solving the E-Waste Problem (StEP) Initiative, launched in 2007, brings together experts from various sectors, including industry, governments, NGOs, and academia, to promote sustainable e-waste management. The initiative aims to foster innovation and collaboration to address the challenges of e-waste management and facilitate the transition to a circular economy. Another notable example is the European Union's Waste Electrical and Electronic Equipment (WEEE) Directive, which was adopted in 2003. This directive mandates EU member states to prevent the generation of e-waste and to promote its reuse, recycling, and recovery. It also requires producers to be responsible for the collection and disposal of their products after the end of their useful lives. In the United States, the Electronic Waste Recycling Act was passed in 2003, which requires certain manufacturers to finance and manage the collection and recycling of their electronic products at the end of their useful lives. The act aims to reduce the amount of electronic waste in landfills and promote sustainable e-waste management. Countries like Japan have also taken steps to promote proper e-waste treatment and recycling to achieve a more sustainable society. The Act on the Promotion of Effective Utilization of Resources, passed in 2001, aims to promote the proper treatment and recycling of e-waste and other resources. The African Circular Economy Alliance was launched in 2020, which seeks to accelerate the transition to a circular economy in Africa, including the sustainable management of e-waste. The alliance aims to promote sustainable economic growth and address environmental challenges by promoting the principles of the circular economy.

5.8 Case Studies

It was reported that a series of workshops were conducted by the researchers in this study, with Finnish electronics manufacturers utilizing the circularity deck. The selected companies were part of a research project that focused on sustainable electronics manufacturing in Finland, indicating their progressive approach to the subject, and offering the potential to yield more positive results compared to the industry as a whole. Due to COVID-related restrictions, the workshops had to be conducted online. Attendees at the workshops consisted of 2–5 members from each company, primarily with backgrounds in engineering/product design and sustainability/marketing. Researchers from participating institutions were also in attendance. At the beginning of the workshops, an introduction to the circularity deck was provided.

An overview of the current state of the circular economy techniques used by Finnish electronics manufacturing enterprises is provided by the bar chart in Fig. 5.5. Only 25% of the enterprises, as seen in the graph, have adopted circular economy initiatives. The "Narrow" category, which focuses on lowering the use of raw materials and comprises 50% of all applied strategies, has the widest adoption of strategies. However, just 13% of the companies have addressed "Near" category tactics, such as recycling electronic devices and using recycled materials, which have gotten little attention. Yet, 60% of the participating businesses have stated that they plan to deal with tactics in this category in the future. An average of 36% of the businesses have circular economy implementation plans for the future. The other 39% of the strategies evaluated, which primarily fell under the "Slow," "Regenerate," and "Inform" classifications, where the share of irrelevant strategies ranged from 39 to 55%, were declared irrelevant for the participating companies and products. Below the graphic is a thorough explanation of each category.

The adoption of circular economy strategies by participating companies, categorizing them into three groups: those already addressed, those yet to be addressed, and those deemed irrelevant. The total number of strategies considered is presented on the left side, while a breakdown by category is displayed on the right side. This assessment offers insights into the companies' current practices and areas where further attention and improvements are needed to promote a more sustainable circular economy approach [32]

According to the report, companies in Finland's electronics sector are eager to improve their circular economy policies by putting various additional measures into practice. Only 25% of cases at this time have used circular economy tactics. The "light-weighting" strategy, which belongs to the "Narrow" category and is the most

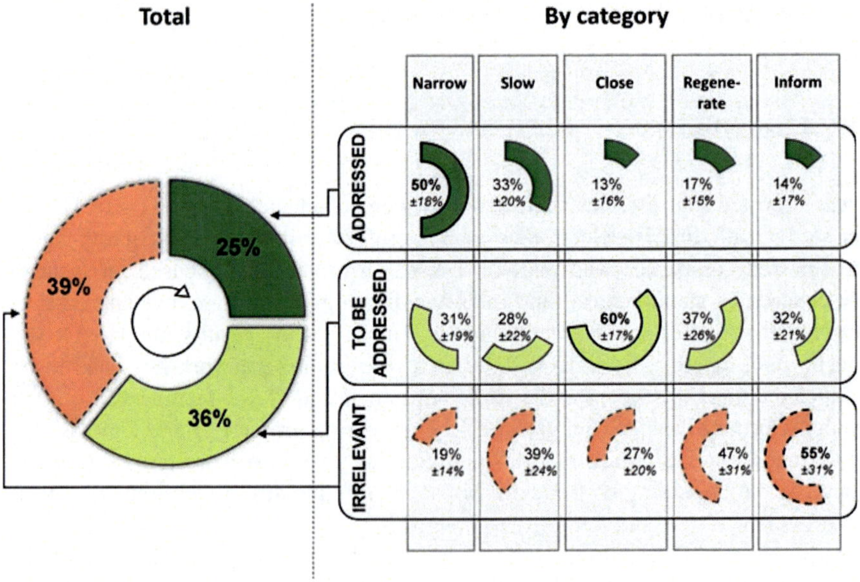

Fig. 5.5 Assessing the circularity practices adopted by companies as list ed in [32]

frequently used, is dedicated to making smaller, thinner, and lighter items. Its acceptance rate is 50%. Additionally, in 33% of the cases, techniques aimed at producing long-lasting, durable goods, or the "Slow" category, were put into practice. The "Close" category, which emphasizes recycling end-of-life products and employing recycled materials, has the most room for growth. Most businesses strive to create items from just one or a few recyclable components that can be used again for primary recycling, as well as to utilise recycled materials in their manufacturing procedures [32].

5.8.1 Business Models Scenario and Considerations

Achieving a circular economy in the electronics industry revolves around the concept of capturing remaining value or utility. New business models are emerging that prioritize access over ownership, allowing manufacturers to maintain ownership and responsibility of their products. This approach has been successful in achieving high rates of recovery and reuse of products like modems and is now being applied to smartphones and laptops as well. By adopting these new business models, companies can not only increase the value of their products, but also foster new relationships with their customers while promoting the continued use of valuable resources. A product's residual value is determined by the functional value that it continues to have over time and by how customers view that value. The value and utility of used electronics can be affected by a number of variables, including product design, refurbishing technology, the rate of technological innovation, logistics, user perception, and the number of items on the market. A thorough comprehension of residual value offers the opportunity to execute systemic change and rethink the interaction between people and technology equipment.

5.9 Circularity in Transportation

Circular mobility systems are multi-modal, providing a diverse range of transportation options that cater to the varied needs of cities and their residents. This includes integrating public transportation with on-demand services for flexible last-mile solutions. New technologies like electric-powered, shared, and automated transportation are gaining popularity in urban areas, enhancing sustainable mobility options.

In the quest for a circular economy in the automotive sector, vital considerations include optimizing vehicle lifespans and finding a delicate equilibrium between environmental impacts. The implementation of extended producer responsibility becomes paramount, as it requires products to be returned for purposes of reuse, recycling, or remanufacturing. Challenges emerge from conflicting circular approaches, the necessity for flexible policies, and the integration of software to support extended lifespans and maintenance. Careful management of trade-offs between recycling and

Table 5.5 Recycling strategies of individual parts of dismantled car waste

Car individual parts	Recycling strategies	Challenges	GHG emissions
Engine	Remanufacturing, component reuse	Complex disassembly, contamination	Emission from combustion
Battery (electric cars)	Direct reuse, refurbishment, recycling	Fire hazards, hazardous contaminants	GHG from battery production, recycling
Tires	Retreading, recycling	Limited end-of-life recycling infrastructure	Emissions from tire production
Metals (e.g., steel)	Melting, refining, shredding	Energy-intensive recycling process	GHG from melting and refining process
Glass	Crushing, melting	Contamination, separating different types of glass	GHG from glass production
Plastics	Shredding, melting	Multiple plastic types, contamination	GHG from plastic production
Electronics	Component reuse, precious metal recovery	E-waste management, complex disassembly	GHG from electronics production
Interior components	Reconditioning, refurbishment	Difficult to separate materials	GHG from production processes

reuse is imperative. Additionally, there are existing data gaps concerning the fate and exportation of vehicle waste, while environmental concerns surround waste electric vehicle batteries. To drive successful circular practices and overcome these obstacles, effective global policies and regulations for waste management are indispensable (Table 5.5).

5.10 Conclusion

Electronic waste is an ever-growing problem that requires immediate attention to ensure that it does not cause further harm to the environment. This chapter provides a comprehensive understanding of electronic waste generation, classification, and recycling strategies. It sheds light on the various stages involved in the recycling process, including collection, sorting, dismantling, and advanced recycling techniques. Moreover, the chapter presents alternate materials and solutions, such as the shared economy model, products-as-a-service model, and product ownership model, that can be used to manage electronic waste sustainably. In addition to this, the chapter explores the potential of organic electronics and IT-enabled electronics as alternatives to traditional electronic devices, which can reduce the generation of electronic waste. It discusses the advantages of organic field effect transistors, photovoltaics, memory devices, and LEDs, which can help reduce the environmental impact of electronic devices. Furthermore, the chapter provides insights into global initiatives and policies related to electronic waste management, which can help guide policymakers, industry professionals, and researchers in developing sustainable solutions for managing electronic waste. The chapter concludes with case studies that consider business models and their implications for sustainable e-waste management. By providing valuable insights into the challenges associated with electronic waste management, this chapter serves as a valuable resource for policymakers, industry professionals, and researchers seeking to address these challenges and develop sustainable solutions for managing electronic waste.

1. What is the primary goal of circular practices in e-waste management and transportation?
   (a) To increase landfill space
   (b) To reduce resource consumption and waste
   (c) To promote single-use products
   (d) To encourage incineration of e-waste
2. Which of the following is NOT a circular practice in e-waste management?
   (a) Recycling electronic components for reuse
   (b) Disposing of e-waste in landfills
   (c) Refurbishing electronic devices for resale
   (d) Remanufacturing electronic products
3. How does circular transportation contribute to e-waste management?
   (a) It increases transportation costs
   (b) It allows for more e-waste exports
   (c) It minimizes the environmental impact of transportation
   (d) It encourages dumping of e-waste in oceans
4. Which of the following strategies can be considered a circular approach to e-waste management?
   (a) Exporting e-waste to developing countries
   (b) Repairing and reusing electronic devices
   (c) Dumping e-waste in open spaces
   (d) Incinerating e-waste
5. Which circular practice in e-waste management can help reduce greenhouse gas emissions?
   (a) Landfilling e-waste
   (b) Recycling electronic components
   (c) Exporting e-waste to foreign countries
   (d) Burning e-waste

6. What is the key advantage of adopting circular practices in e-waste management?
   (a) Increased environmental pollution
   (b) Higher resource consumption
   (c) Greater cost savings and sustainability
   (d) Lower product lifespan
7. Which circular approach involves disassembling used products for reusable parts?
   (a) Recycling
   (b) Landfilling
   (c) Remanufacturing
   (d) Incineration
8. How does circular transportation contribute to reducing e-waste?
   (a) By increasing waste generation
   (b) By promoting illegal dumping
   (c) By minimizing the distance traveled for recycling
   (d) By encouraging overseas e-waste disposal
9. Which of the following is a circular practice for e-waste management that prioritizes product longevity?
   (a) Planned obsolescence
   (b) Reusing and refurbishing electronic devices
   (c) Single-use electronic products
   (d) Rapid disposal of electronic waste
10. Circular practices in e-waste management and transportation aim to achieve:
   (a) Increased environmental pollution
   (b) Greater resource consumption
   (c) Sustainable and efficient resource utilization
   (d) Escalating electronic waste generation

Answers:

1. (b) To reduce resource consumption and waste
2. (b) Disposing of e-waste in landfills
3. (c) It minimizes the environmental impact of transportation
4. (b) Repairing and reusing electronic devices
5. (b) Recycling electronic components
6. (c) Greater cost savings and sustainability
7. (c) Remanufacturing
8. (c) By minimizing the distance traveled for recycling
9. (b) Reusing and refurbishing electronic devices
10. (c) Sustainable and efficient resource utilization

References

1. Palaniandy, P.; Yusoff, M.S.; Wang, L.K.; Wang, M.-H.S. Electronic and Electrical Equipment Waste Disposal. **2022**, 115–161, doi:https://doi.org/10.1007/978-3-030-96989-9_3.
2. Ankit; Saha, L.; Kumar, V.; Tiwari, J.; Sweta; Rawat, S.; Singh, J.; Bauddh, K. Electronic Waste and Their Leachates Impact on Human Health and Environment: Global Ecological Threat and Management. *Environ Technol Innov* **2021**, *24*, 102049, doi:https://doi.org/10.1016/J.ETI.2021.102049.
3. Compagnoni, M. Is Extended Producer Responsibility Living up to Expectations? A Systematic Literature Review Focusing on Electronic Waste. *J Clean Prod* **2022**, *367*, 133101, doi:https://doi.org/10.1016/J.JCLEPRO.2022.133101.
4. Esenduran, G.; Atasu, A.; Van Wassenhove, L.N. Valuable E-Waste: Implications for Extended Producer Responsibility. https://doi.org/10.1080/24725854.2018.1515515 **2019**, *51*, 382–396, doi:https://doi.org/10.1080/24725854.2018.1515515.
5. Atasu, A.; Subramanian, R. Extended Producer Responsibility for E-Waste: Individual or Collective Producer Responsibility? *Prod Oper Manag* **2012**, *21*, 1042–1059, doi:https://doi.org/10.1111/J.1937-5956.2012.01327.X.
6. Pathak, P.; Srivastava, R.R.; Ojasvi Environmental Management of E-Waste. *Electronic Waste Management and Treatment Technology* **2019**, 103–132, doi:https://doi.org/10.1016/B978-0-12-816190-6.00005-4.
7. Pandey, J.; Kaushal, R.K.; Shukla, S.P. Behavioural and Structural Analysis of Stakeholders in E-Waste Management. *Mater Today Proc* **2022**, *51*, 416–421, doi:https://doi.org/10.1016/J.MATPR.2021.05.566.
8. Leclerc, S.H.; Badami, M.G. Extended Producer Responsibility for E-Waste Management: Policy Drivers and Challenges. *J Clean Prod* **2020**, *251*, 119657, doi:https://doi.org/10.1016/J.JCLEPRO.2019.119657.
9. Kawecki, D.; Wu, Q.; Gonçalves, J.S.V.; Nowack, B. Polymer-Specific Dynamic Probabilistic Material Flow Analysis of Seven Polymers in Europe from 1950 to 2016. *Resour Conserv Recycl* **2021**, *173*, 105733, doi:https://doi.org/10.1016/J.RESCONREC.2021.105733.
10. Bubnova, O.; Crispin, X. Towards Polymer-Based Organic Thermoelectric Generators. *Energy Environ Sci* **2012**, *5*, 9345–9362, doi:https://doi.org/10.1039/C2EE22777K.
11. Shevchenko, T.; Laitala, K.; Danko, Y. Understanding Consumer E-Waste Recycling Behavior: Introducing a New Economic Incentive to Increase the Collection Rates. *Sustainability 2019, Vol. 11, Page 2656* **2019**, *11*, 2656, doi:https://doi.org/10.3390/SU11092656.
12. Khetriwal, D.S.; Kraeuchi, P.; Widmer, R. Producer Responsibility for E-Waste Management: Key Issues for Consideration – Learning from the Swiss Experience. *J Environ Manage* **2009**, *90*, 153–165, doi:https://doi.org/10.1016/J.JENVMAN.2007.08.019.
13. Murthy, V.; Ramakrishna, S. A Review on Global E-Waste Management: Urban Mining towards a Sustainable Future and Circular Economy. *Sustainability 2022, Vol. 14, Page 647* **2022**, *14*, 647, doi:https://doi.org/10.3390/SU14020647.
14. Arya, S.; Rautela, R.; Chavan, D.; Kumar, S. Evaluation of Soil Contamination Due to Crude E-Waste Recycling Activities in the Capital City of India. *Process Safety and Environmental Protection* **2021**, *152*, 641–653, doi:https://doi.org/10.1016/J.PSEP.2021.07.001.
15. Saveyn, H.; Eder, P.; Garbarino, E.; Muchova, L.; Hjelmar, O.; van der Sloot, H.; Comans, R.; van Zomeren, A.; Hyks, J.; Oberender, A. Study on Methodological Aspects Regarding Limit Values for Pollutants in Aggregates in the Context of the Possible Development of End-of-Waste Criteria under the EU Waste Framework Directive., doi:https://doi.org/10.2791/1125.
16. Farshad, A.; Gholami, H.; Farzadkia, M.; Mirkazemi, R.; Kermani, M. The Safety of Non-Incineration Waste Disposal Devices in Four Hospitals of Tehran. *Int J Occup Environ Health* **2014**, *20*, 258–263, doi:https://doi.org/10.1179/2049396714Y.0000000072.

17. Dubai Municipality Promotes E-Waste Awareness in Educational Sector Available online: https://www.dm.gov.ae/2022/03/30/dubai-municipality-promotes-e-waste-awareness-in-educational-sector/ (accessed on 22 July 2023).
18. Jaidev, K.; Biswal, M.; Mohanty, S.; Nayak, S.K. Sustainable Waste Management of Engineering Plastics Generated from E-Waste: A Critical Evaluation of Mechanical, Thermal and Morphological Properties. *J Polym Environ* **2021**, *29*, 1763–1776, doi:https://doi.org/10.1007/S10924-020-01998-Z/METRICS.
19. Laszlo, R.; Holonec, R.; Copîndean, R.; Dragan, F. Sorting System for E-Waste Recycling Using Contour Vision Sensors. *Proceedings of 2019 8th International Conference on Modern Power Systems, MPS 2019* **2019**, doi:https://doi.org/10.1109/MPS.2019.8759739.
20. Rochman, F.F.; Ashton, W.S.; Wiharjo, M.G.M. E-Waste, Money and Power: Mapping Electronic Waste Flows in Yogyakarta, Indonesia. *Environ Dev* **2017**, *24*, 1–8, doi:https://doi.org/10.1016/J.ENVDEV.2017.02.002.
21. Ali, L.; Sivaramakrishnan, K.; Kuttiyathil, M.S.; Chandrasekaran, V.; Ahmed, O.H.; Al-Harahsheh, M.; Altarawneh, M. Prediction of Thermogravimetric Data in Bromine Captured from Brominated Flame Retardants (BFRs) in e-Waste Treatment Using Machine Learning Approaches. *J Chem Inf Model* **2023**, *63*, 2305–2320, doi:https://doi.org/10.1021/ACS.JCIM.3C00183.
22. Shah, S.; Houda, M.; Khan, S.; Althoey, F.; Abuhussain, M.; Abuhussain, M.A.; Ali, M.; Alaskar, A.; Javed, M.F. Mechanical Behaviour of E-Waste Aggregate Concrete Using a Novel Machine Learning Algorithm: Multi Expression Programming (MEP). *Journal of Materials Research and Technology* **2023**, *25*, 5720–5740, doi:https://doi.org/10.1016/J.JMRT.2023.07.041.
23. Jana, R.K.; Ghosh, I.; Das, D.; Dutta, A. Determinants of Electronic Waste Generation in Bitcoin Network: Evidence from the Machine Learning Approach. *Technol Forecast Soc Change* **2021**, *173*, 121101, doi:https://doi.org/10.1016/J.TECHFORE.2021.121101.
24. Chen, D.; Liu, R.; Lin, Q.; Ma, S.; Li, G.; Yu, Y.; Zhang, C.; An, T. Volatile Organic Compounds in an E-Waste Dismantling Region: From Spatial-Seasonal Variation to Human Health Impact. *Chemosphere* **2021**, *275*, 130022, doi:https://doi.org/10.1016/J.CHEMOSPHERE.2021.130022.
25. Ge, X.; Ma, S.; Zhang, X.; Yang, Y.; Li, G.; Yu, Y. Halogenated and Organophosphorous Flame Retardants in Surface Soils from an E-Waste Dismantling Park and Its Surrounding Area: Distributions, Sources, and Human Health Risks. *Environ Int* **2020**, *139*, 105741, doi:https://doi.org/10.1016/J.ENVINT.2020.105741.
26. Li, X.; Wu, Y.; Tan, Z. An Overview on Bioremediation Technologies for Soil Pollution in E-Waste Dismantling Areas. *J Environ Chem Eng* **2022**, *10*, 107839, doi:https://doi.org/10.1016/J.JECE.2022.107839.
27. Verbeeck, K.; Buelens, L.C.; Galvita, V. V.; Marin, G.B.; Van Geem, K.M.; Rabaey, K. Upgrading the Value of Anaerobic Digestion via Chemical Production from Grid Injected Biomethane. *Energy Environ Sci* **2018**, *11*, 1788–1802, doi:https://doi.org/10.1039/C8EE01059E.
28. 29. Li, K.; Xu, Z. A Review of Current Progress of Supercritical Fluid Technologies for E-Waste Treatment. *J Clean Prod* **2019**, *227*, 794–809, doi:https://doi.org/10.1016/J.JCLEPRO.2019.04.104.
29. Umicore l A Global Materials Technology and Recycling Group Available online: https://www.umicore.com/en/ (accessed on 22 July 2023).
30. Zhu, M.; Jia, C.; Wang, Y.; Fang, Z.; Dai, J.; Xu, L.; Huang, D.; Wu, J.; Li, Y.; Song, J.; et al. Isotropic Paper Directly from Anisotropic Wood: Top-Down Green Transparent Substrate Toward Biodegradable Electronics. *ACS Appl Mater Interfaces* **2018**, *10*, 28566–28571, doi:https://doi.org/10.1021/ACSAMI.8B08055.

31. Kim, E.; Lee, J.; Kim, D.; Lee, K.E.; Han, S.S.; Lim, N.; Kang, J.; Park, C.G.; Kim, K. Solvent-Responsive Polymernanocapsules with Controlled Permeability: Encapsulation and Release of a Fluorescent Dye by Swelling and Deswelling. *Chemical Communications* **2009**, 1472–1474, doi:https://doi.org/10.1039/B823110A.
32. Deviatkin, I.; Rousu, S.; Ghoreishi, M.; Naji Nassajfar, M.; Horttanainen, M.; Leminen, V. Implementation of Circular Economy Strategies within the Electronics Sector: Insights from Finnish Companies. *Sustainability 2022, Vol. 14, Page 3268* **2022**, *14*, 3268, doi:https://doi.org/10.3390/SU14063268.

Chapter 6
Circular Approaches in Fashion Industries and Building Materials

Abstract The adoption of circular approaches has become increasingly critical in the fashion and construction industries to address sustainability challenges and promote resource efficiency. This chapter explores circularity in the fashion sector, highlighting circular design principles, biomimicry-inspired and intelligent materials, zero-waste pattern cutting, biodegradable materials, and textile-to-textile recycling. It further delves into the circularity of specific materials in textile fashion, including cellulose, polyester, polyurethane, polyolefins, polyamide, and polyacrylics. The circular models adopted in the fashion industries are examined, with a focus on remanufacturing and modular design. In the construction domain, circular approaches are applied to building material selection, with an analysis of materials like concrete, steel, wood, and other alternatives. The critical parameters impacting building materials, such as humidity, diffusion, and vaporization, are explored to optimize material performance and reduce environmental impact. The chapter also presents global initiatives driving sustainable practices in both fashion and construction industries. By embracing circularity, the fashion and construction sectors can play a pivotal role in promoting environmental stewardship and fostering a regenerative economy.

Keywords Nature-inspired materials · Digital pattern cutting · Sustainable polymers · Concrete's carbon footprint · Water sorption capacity

6.1 Circular Fashion Economy

Circular fashion, based on principles of waste and pollution reduction, extended product use, and natural system regeneration, minimizes waste and resource use. Environmental benefits include waste and pollution reduction through innovative design, repair, and recycling [1, 2]. Circular fashion supports biodiversity and ecosystem health by using sustainable materials, reducing resource-intensive materials, and addressing microplastic pollution. Innovations driving circular fashion

S. Ramakrishna and B. Ramasubramanian, *Handbook of Materials Circular Economy*,
https://doi.org/10.1007/978-981-97-0589-4_6

include sustainable materials, eco-friendly production processes, digital technologies for traceability, and new business models [3, 4]. Clothing rental services, repair and resale platforms, collaborative consumption, zero-waste and upcycling designs, and made-to-order production contribute to circularity. Embracing circular fashion can significantly reduce the fashion industry's environmental impact, supporting environmental conservation and restoration goals. It offers a sustainable alternative to traditional practices, contributing to a greener future. For instance, Patagonia exemplifies a circular fashion brand, creating durable, high-quality products and implementing the "Worn Wear" program, which repairs and resells used clothing [5].

Textile waste in the US, mainly discarded apparel, is increasing over time, comprising 5.3% (13.2 MT) of total MSW in 2010, and growing to 6.3% (16.9 MT) in 2017 (Fig. 6.1a). Approximately 85% of textiles end up in landfills, with the US having the highest landfilling rate (29.3 kg/ca in 2016) among leading economies. Encouraging recycling technologies becomes crucial to address this issue, as incineration and recycling gain popularity in textile waste management (Fig. 6.1c and d) [6].

6.2 Circular Design Principles in Fashion

Circular design principles in fashion aim to address the environmental impacts of the fashion industry by adopting a holistic approach that considers the entire lifecycle of products. Material selection plays a crucial role in achieving circularity. By opting for renewable fibers like organic cotton, designers can reduce the use of harmful chemicals and promote more sustainable agricultural practices [7]. Additionally, choosing biodegradable fibers ensures that garments can naturally decompose at the end of their life, reducing the burden on landfills [8, 9].

In the current linear fashion chain, resources are mainly either fossil-based or of renewable origin, with synthetic polymers from fossil resources and natural polymers like cellulose from renewable resources. In a future scenario, it may be possible to make synthetic polymers from renewables. In the circular fashion chain, resources come from renewable sources and post-consumer or post-industrial residues in the form of yarns or fabrics [8]. The post-consumer stream poses challenges due to contamination and decreased fiber quality. To maintain quality, virgin renewable resources are required in combination with recycling. Monomer, polymer, and fiber recycling each follow specific pathways and re-enter the textile production cycle at different levels. The goal is to preserve the material structure and minimize processing to promote sustainability. The figure illustrates the relationship between resources and processes involved in producing garments in the fashion industry [9]. The upper part of the Fig. 6.2 represents the linear route, where resources mainly originate from fossil-based and renewable sources. Fossil-based resources are used to produce synthetic polymers, essential components for various synthetic fibers in textiles. Renewable resources contribute natural polymers like cellulose, obtained from wood,

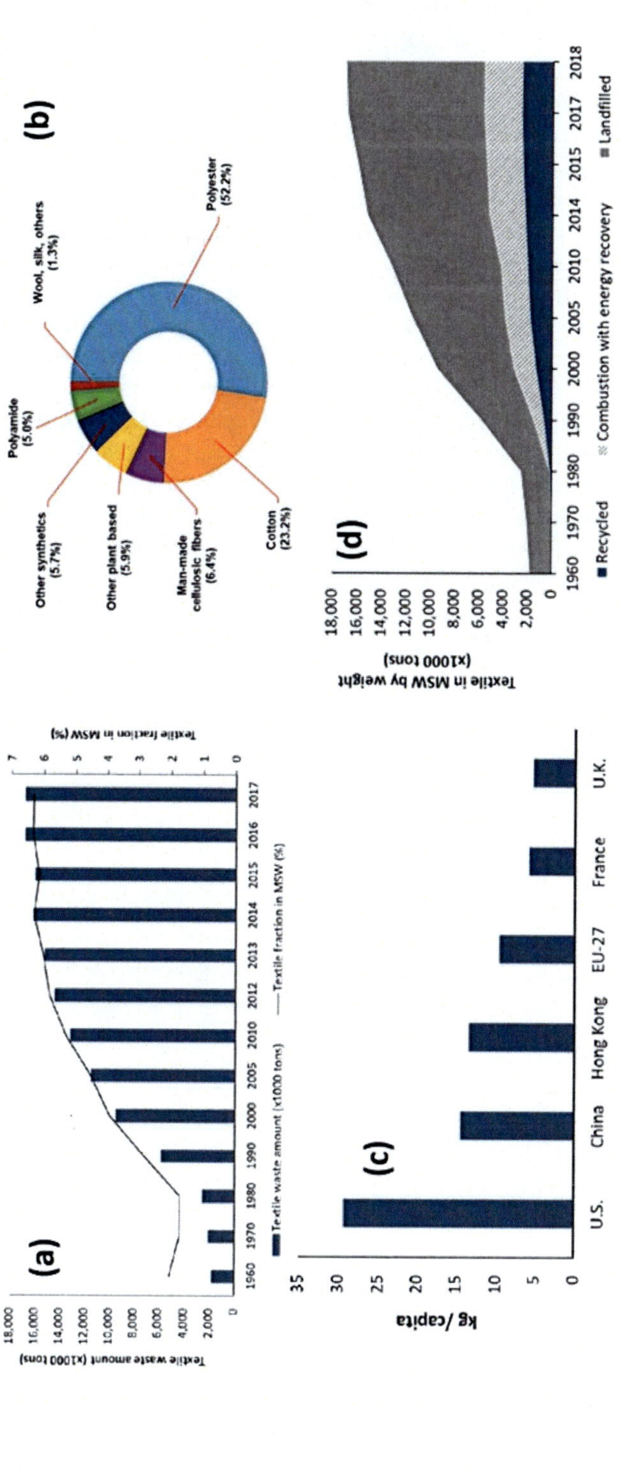

Fig. 6.1 **a** Textile Trash in the USA, **b** Global Fiber Production Share 2019, **c** Landfilled Textiles in 2016 (kg/ca) and **d** USA's Textile Waste Solutions [6]

which is then transformed into regenerated cellulose fibers through extrusion spinning [9]. In contrast, the lower part of the Fig. 6.2 depicts the circular route, driven by sustainability and waste reduction principles. Resources for the circular route come primarily from renewable sources and post-consumer or post-industrial residues in the form of yarns or fabrics. This circular approach aims to minimize waste and maximize resource efficiency by reusing, recycling, and reducing waste, creating a closed-loop system to promote environmental sustainability in the fashion industry. However, the post-consumer stream presents challenges, including contamination and reduced fiber quality due to washing and wearing. To maintain quality in the circular route, a combination of recycling and the incorporation of virgin renewable resources may be necessary, particularly in polymer and fiber recycling processes [11].

Recycling is a key aspect of circularity, and this applies to both synthetic and natural fibers. Prioritizing recycled polyester (rPET) reduces the demand for virgin polyester, which requires significant energy and resources to produce. Developing mono-material designs is beneficial for recycling, as it simplifies the separation process and enables more efficient reuse of materials. Furthermore, exploring innovative recycling technologies like chemical recycling offers the potential to break down polyester into its raw materials, enabling the creation of new fibers with minimal degradation in quality [12]. Additional, Wool, a natural and biodegradable fiber, can be made more circular through regenerative grazing practices. These practices focus

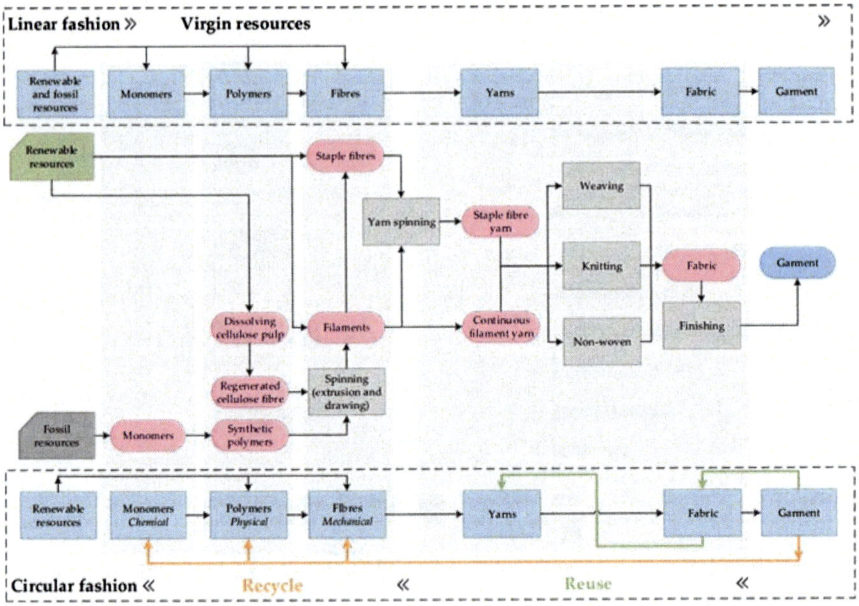

Fig. 6.2 Relationship between resources and processes involved in producing garments in the fashion industry [10]

on improving soil health and biodiversity, making wool production more sustainable. Designing durable and timeless wool garments enhances their longevity, reducing the frequency of replacements and, consequently, overall consumption. Implementing take-back programs for wool products allows for the recovery of valuable fibers and materials, promoting circularity in the fashion industry [13, 14].

In similar way, Lyocell, commonly known as Tencel, is produced from sustainably managed wood sources. This renewable and biodegradable fiber offers an eco-friendlier alternative to conventional textiles. Closed-loop manufacturing processes in lyocell production minimize waste by efficiently recycling solvents and reducing water consumption. Circular business models, such as garment rental or leasing, encourage prolonged product use and discourage disposability, contributing to a more sustainable fashion ecosystem [15]. However, raw materials like leather poses unique challenges in circularity due to its complex production process. Alternatively, vegetable-tanned or chrome-free leather, which reduces the environmental impact of traditional leather tanning processes can be used. Promoting leather recycling or repurposing initiatives can extend the life of leather products and minimize waste [15, 16]. Exploring alternative materials, such as mushroom or pineapple leather, presents exciting opportunities for more sustainable leather alternatives. Below are the five sustainable circular design principles:

I. Biomimicry-inspired Materials: Drawing inspiration from nature's efficient design strategies, biomimicry-inspired materials like spider silk-inspired fibers are utilized to enhance durability and sustainability while reducing reliance on non-renewable resources [16].
II. Zero-Waste Pattern Cutting: Implementing precise pattern cutting techniques minimizes fabric waste during garment production. Utilizing computer-aided design (CAD) and 3D modelling optimizes pattern layouts, ensuring maximum material utilization and reducing excess fabric scraps.
III. Biodegradable and Compostable Materials: Prioritizing the use of biodegradable and compostable textiles ensures that garments can naturally break down at the end of their life cycle. Exploring innovative materials like algae-based fabrics or cellulose-based fibers derived from agricultural waste promotes a closed-loop system where textiles can return to the earth harmlessly [17].
IV. Modular Design and Remanufacturing: Designing garments in modular components facilitates easy disassembly, repair, or upgrades. The implementation of snap-on buttons, zippers, or hook-and-loop fasteners enables garment remanufacturing, extending their usability and reducing the need for new purchases.
V. Textile-to-Textile Recycling: Integrating textile-to-textile recycling processes recovers fibers from discarded garments, allowing the creation of new textiles without degradation in quality. Utilizing mechanical, chemical, or enzymatic recycling technologies breaks down fabrics into their constituent fibers, enabling the production of new garments from post-consumer textiles.

6.2.1 Biomimicry-Inspired and Intelligent Materials

Biomimicry-inspired materials emulate nature's efficient design strategies, borrowing structural, functional, and compositional features from living organisms. By understanding and harnessing the sophisticated processes that have evolved over millions of years, scientists and engineers have unlocked novel solutions to real-world challenges. One such example is lotus-inspired superhydrophobic surfaces, which feature microscopic structures that repel water and prevent dirt and contaminants' adhesion [17, 18]. These surfaces find applications in self-cleaning coatings, anti-icing materials, and water-repellent textiles. Spider silk-inspired fibers, known for their strength and flexibility, have led to the development of bioengineered fibers with exceptional tensile strength and elasticity. These fibers are suitable for lightweight and high-performance materials, including bulletproof vests and medical sutures. Additionally, bone-inspired materials mimic the hierarchical structure of bones, resulting in lightweight and durable composites with remarkable load-bearing capabilities, promising revolutionary applications in construction and aerospace industries. Moreover, gecko-inspired adhesives, inspired by geckos' toe pads, enable reversible, strong, and residue-free adhesion [18, 19].

Intelligent materials, also known as smart materials, respond to external stimuli, enabling controlled and predictable changes in their properties. Shape Memory Alloys (SMAs) can "remember" their original shape and revert to it when heated after being deformed. Piezoelectric materials generate electric charges when mechanically stressed, allowing them to convert mechanical energy into electrical energy and vice versa. Electrochromic materials change colour in response to an electric current, offering controllable tinting and light-modulating properties, making them ideal for advanced fashion products that adjust their transparency based on external conditions. Furthermore, thermochromic materials change color with temperature variations, serving as thermal indicators with applications in smart textiles, visual temperature sensors, and novelty items like color-changing costumes. The convergence of biomimicry-inspired and intelligent materials (Fig. 6.3) has led to transformative advancements across industries. Sustainable architecture can leverage biomimicry to optimize energy efficiency and incorporate intelligent materials into smart facades and electric dresses for adaptable insulation [20, 21].

6.2.2 Zero-Waste Pattern Cutting

Zero-waste pattern cutting is a highly technical and sustainable design approach in the fashion industry aimed at minimizing textile waste during garment production. Unlike traditional methods, zero-waste cutting maximizes fabric utilization through precise and strategic placement of pattern pieces, taking into account fabric width, length, and stretch. This requires a deep understanding of garment construction and fabric behavior, as designers carefully plan pattern layouts to avoid leftover

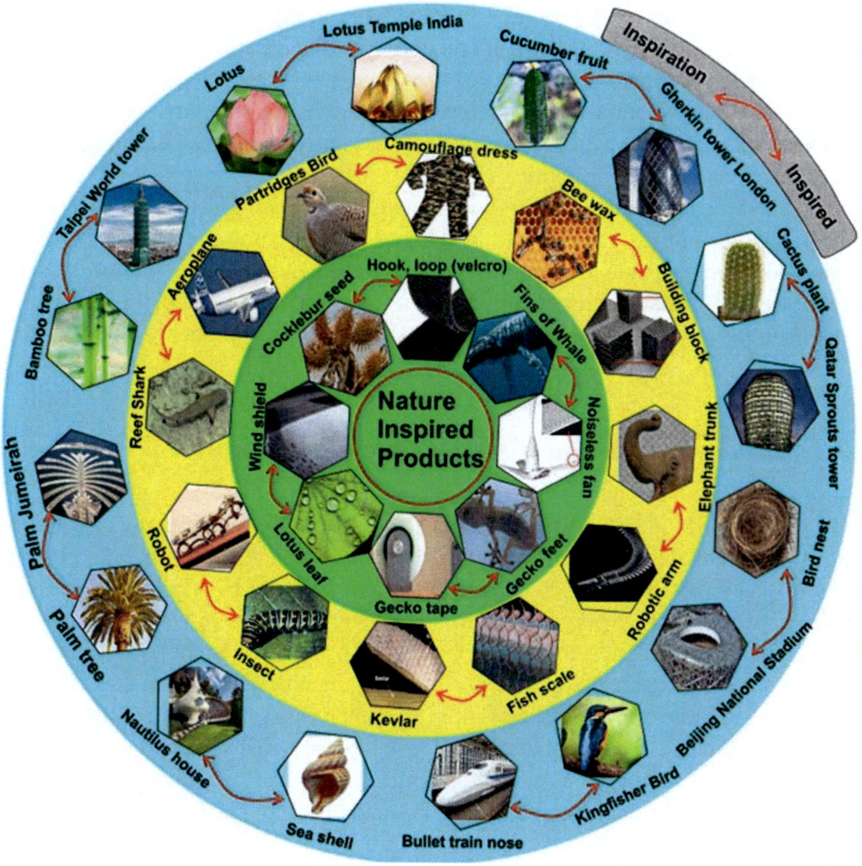

Fig. 6.3 Schematic of nature, bio-inspired, bio-mimetic systems; Green circles represent functional mimetics, yellow circles represent feature mimetics, and cyan circles symbolize world-remarkable architecture inspired by nature [22]

scraps. The methodology incorporates modular design principles, allowing garment components to be reused for multiple styles, reducing fabric waste [23]. Ensuring pattern pieces are cut on the fabric's grainline is vital to maintain garment stability and eliminate off-cuts. Creative seam placement further optimizes fabric usage, enhancing the garment's aesthetics while minimizing scrap. To facilitate efficient cutting, digital technology such as computer-aided design (CAD) software and 3D modelling is employed to simulate pattern layouts and identify potential issues before cutting. These include precise pattern creation and editing, automated grading for different sizes, virtual prototyping to visualize garments on a 3D model, and pattern layout simulation to optimize material usage. Nesting features arrange pattern pieces efficiently on fabric, reducing waste. Advanced CAD software can simulate fabric behavior, aiding in garment design. Measurement tools ensure accuracy, while layer management organizes complex designs. Pattern libraries and export options streamline the process, making CAD and 3D modelling indispensable for sustainable and efficient fashion production [24, 25].

However, challenges and limitations persist, including the complexity of achieving zero-waste patterns, particularly for intricate designs, requiring careful planning and experimentation. Some fabrics may be less amenable to this method, such as those with directional prints or limited stretch, necessitating careful fabric selection. Balancing fabric efficiency with garment fit and comfort can also be challenging. Widespread industry adoption of zero-waste cutting necessitates a shift in practices and mindset, with education and training being essential for designers, manufacturers, and consumers to embrace this sustainable approach fully.

6.2.3 *Biodegradable and Compostable Materials*

In the domain of biodegradable and compostable materials, a wide array of classifications illustrates their adaptability and potential in advancing sustainability. It was observed that (i) bio-based biodegradable polymers, such as PHA (Polyhydroxyalkanoates) produced by microorganisms from plant sugars, exhibit diverse applications in packaging and cosmetics. PLA (Polylactic Acid), derived from renewable sources like corn-starch or sugarcane, serves in packaging, disposable bags, and textiles due to its biodegradable nature and versatility. Another example includes PBS (Polybutylene Succinate), a biodegradable polyester generated by bacteria from plant sugars, finding uses in agricultural films, mulch films, and injection-molded products [19, 21]. Notably, the combination of PHA and PLA into PHA-PLA blends offers improved mechanical properties and a broader range of use in fashion industry. Subsequently, (ii) natural biodegradable materials encompass cellulose-based materials derived from plant fibers, commonly adopted in packaging, textiles, and food coatings. Pectin-based materials, sourced from fruits, can be modified to create biodegradable coatings and films. Additionally, chitosan-based materials, acquired from crustacean shells, demonstrate biodegradability water treatment, and packaging. Furthermore, (iii) synthetic biodegradable polymers, like PBSA (Polybutylene Succinate Adipate), composed of succinic acid and adipic acid, illustrate outstanding biodegradability and extensive usage in packaging and agricultural films. PBAT (Polybutylene Adipate Terephthalate) and PCL (Polycaprolactone), a polyester derived from renewable resources, widely contributes to compostable packaging.

Notably, the category of (iv) hybrid biodegradable materials, such as biodegradable nanocomposites that incorporate natural fillers like cellulose nanofibers into biodegradable polymers to enhance mechanical properties. Additionally, hybrid films and coatings integrate different biodegradable materials, such as chitosan and pectin, in multilayer applications, offering enhanced barrier properties and a wide array of uses in textile packaging. Lastly, exploration of (v) food waste-derived biodegradable materials, like agro-industrial residue-based materials, involves utilization of waste from agriculture and food processing, such as wheat straw, rice husk, or fruit peels, to produce biodegradable materials that contribute to waste reduction and promote circularity. Moreover, coffee grounds bioplastics offer an innovative solution to repurpose coffee waste into biodegradable materials for single-use items and packaging [26] (Table 6.1).

Table 6.1 List of common polymers used in textile fabric and their ease of recycling

Polymer	Chemical formula	Functionality	Ease of recycling	Sustainability rating
Polyester	$(C_{10}H_8O_4)_n$	Durable, wrinkle-resistant, lightweight, versatile	Relatively easy	Moderate
Nylon/polyamide	$(C_{12}H_{22}N_2O_2)_n$	Strong, lightweight, abrasion-resistant, elastic	Difficult	Moderate
Polypropylene	$(C_3H_6)_n$	Lightweight, moisture-wicking, stain-resistant	Challenging	Low to Moderate
Acrylic	$(C_3H_4O_2)_n$	Soft, warm, wool substitute	Limited	Low
Rayon/viscose	$(C_6H_{10}O_5)_n$	Soft, breathable, versatile	Recyclable	Moderate
Lyocell/tencel	$(C_6H_{10}O_5)_n$	Soft, biodegradable, eco-friendly	Recyclable	High
Spandex/elastane	$(C_{12}H_{20}O_2)_n$	Highly elastic, stretch and recovery properties	Difficult	Low
Cotton	$(C_6H_{10}O_5)_n$	Soft, breathable, absorbent	Recyclable	Moderate to High
Wool	$C_nH_{2n}O_nN_n$	Warm, insulating, flame-resistant	Recyclable	High
Silk	$(C_{15}H_{23}N_5O_6)_n$	Luxurious, soft, natural sheen	Recyclable	High
Hemp	$(C_6H_{10}O_5)_n$	Strong, durable, environmentally friendly	Recyclable	High
Linen	$(C_6H_{10}O_5)_n$	Cool, breathable, lustrous	Recyclable	High

High highly sustainable and less carbon footprint, *Low* less sustainable and less carbon footprint, *Moderate* between high and low

6.2.4 Modular Design and Remanufacturing

Modular design in fashion involves creating garments with interchangeable parts, enabling easy assembly and disassembly. It extends product lifespans by allowing consumers to update and modify their clothing, reducing the need for constant new purchases and curbing excessive consumption. This approach aligns well with capsule wardrobes, promoting conscious and thoughtful consumption [27, 28]. Remanufacturing complements this by refurbishing and upcycling used clothing, reducing waste and promoting resource efficiency. It presents economic benefits, like new business models such as fashion rental services, and opportunities for innovation through collaborative efforts. Challenges include creating durable and aesthetically appealing modular components and raising consumer awareness of circular fashion. Modular Design for Circularity in Fashion includes detachable sleeves, mix-and-match components, snap-on accessories, zip-in layers, convertible dresses, multi-functional bags, convertible footwear, multi-way scarves, layered skirts, and collapsible hats, offering versatile and sustainable options for consumers to customize their clothing and accessories, reduce waste, and create multiple outfits from a few pieces [27–29].

6.2.5 Textile-to-Textile Recycling

Textile to textile recycling transforms used textiles into new fibers for creating new garments and products. The process includes collection, sorting, cleaning, shredding, fiber production, weaving/knitting, and manufacturing. Used textiles are collected, sorted based on material type and condition, cleaned, shredded into fibers, and then woven or knitted into new fabrics for manufacturing new textile items [30, 31]. In 2020, the fashion brand Monki released the world's first collection using fibers produced by the Green Machine, showcasing the successful implementation of this sustainable recycling technology in the fashion industry. The Green Machine is an innovative closed-loop system that utilizes water, heat, pressure, and green chemicals to fully separate and recycle cotton and polyester blends into new fibers [32] (Table 6.2).

6.3 Materials Circularity in Textile Fashion

In this section, we explore the applicability of three different recycling approaches for different polymer types commonly used in the textile industry: cellulose, polyester, polyamide, polyurethane, polyolefins, polyacrylics, polypropylene, polylactic acid (PLA), polycarbonate, and polyethylene terephthalate (PET). Understanding the

Table 6.2 Different recycling techniques and suitable polymers

Recycling technique	Ease	Recycled polymer	Maintenance	Cost
Mechanical recycling	Moderate	Polyester, cotton	Regular cleaning and sorting	Moderate
Chemical recycling	Challenging	Polyester, nylon	Complex chemical processes	High
Biological recycling	Promising	Cellulose (cotton)	Biotechnological advancements	Medium to High
Monomer recycling	Complex	Nylon	Precise separation and purification	High
Upcycling	Moderate	Various	Design and creative efforts	Variable
Mechanical fiber recycling	Moderate	Polyester, polypropylene	Sorting and cleaning	Moderate

recyclability and suitability of these diverse polymers is crucial for advancing materials circularity in the textile fashion sector.

6.3.1 Cellulose

Cellulose is a natural polymer that makes up the primary structural component of plant cell walls, and it is the most abundant organic compound on Earth. In the context of textiles, cellulose-based fibers are derived from plant sources such as cotton, flax, hemp, and bamboo. Cellulose recycling can be accomplished using mechanical or physical methods. Mechanical recycling results in shorter fibers suitable for nonwovens and flock applications. On the contrary, physical recycling focuses on polymer recycling to create regenerated cellulose fibers [33, 34]. Wood is typically the primary resource for this process, but post-consumer textiles are also viable. Cotton fibers, which are mostly composed of cellulose, are particularly well-suited for the viscose and lyocell processes, employed by several entities in a limited capacity. Cellulose fiber recycling faces challenges, including contamination, shorter fiber length, complex chemical treatment, difficulties in post-consumer collection, scaling up small-scale processes, and managing environmental impacts [33, 34] (Table 6.3).

6.3.2 Polyester

Polyester, particularly polyethylene terephthalate (PET), is extensively utilized in the apparel industry, making PET recycling crucial as its usage in textiles continues to rise. In textile applications, physical and chemical recycling are the most viable

Table 6.3 List of cellulose recycling companies

Company name	Resource from which cellulose derived	Country
Lenzing AG	Wood pulp	Austria
Aditya Birla Group	Wood pulp	India
Sateri	Wood pulp	China
Tangshan Sanyou	Wood pulp	China
Fulida	Wood pulp	China
Kelheim Fibres	Wood pulp	Germany
Bracell	Wood pulp	Brazil
GrupoSniace	Eucalyptus wood	Spain
Frankenhuis	Post-consumer textile waste (cotton)	The Netherlands
Wolkat	Post-consumer textile waste (cotton)	The Netherlands and Morocco
Belda Llorens	Cotton waste a.o	Spain
Geetanjali Woollens	Cotton waste a.o	India
Ferre	Cotton waste	Spain
Velener Textil GmbH	Post-industrial cotton yarns	Germany
Renewcell	High-content cellulose waste (cotton, regenerated cellulose)	Sweden
Evrnu	High-content cellulose waste (cotton, regenerated cellulose)	USA
Infinited Fibre	High-content cellulose waste (cotton, regenerated cellulose)	Finland
Aalto University	High-content cellulose waste (cotton, regenerated cellulose)	Finland
SaXcell	High-content cellulose waste (cotton, regenerated cellulose)	The Netherlands

options for PET, while mechanical recycling presents more challenges. Physical recycling involves melting PET, a thermoplastic material, and re-spinning it into fibers. Transparent post-consumer bottles are commonly utilized to produce high-quality recycled PET suitable for yarn production. However, recycling coloured bottles, trays, films, and PET recovered from the ocean proves to be more complex, necessitating potential employment of chemical recycling methods, Fig. 6.4 [36, 37]. Chemical recycling efficiently breaks down polyester molecules into smaller fragments, enhancing the removal of contaminants compared to mechanical and physical recycling. This method is particularly well-suited for PET fiber production, offering promising possibilities for sustainable textile manufacturing.

Companies such as Velener Textil GmbH and Cumapol are engaged in PET recycling, with Velener Textil producing PET-woven fabric from PET bottles and Cumapol developing CuRe Technology to transform colored PET into transparent PET granulate. The chemical recycling of PET is achieved through various innovative

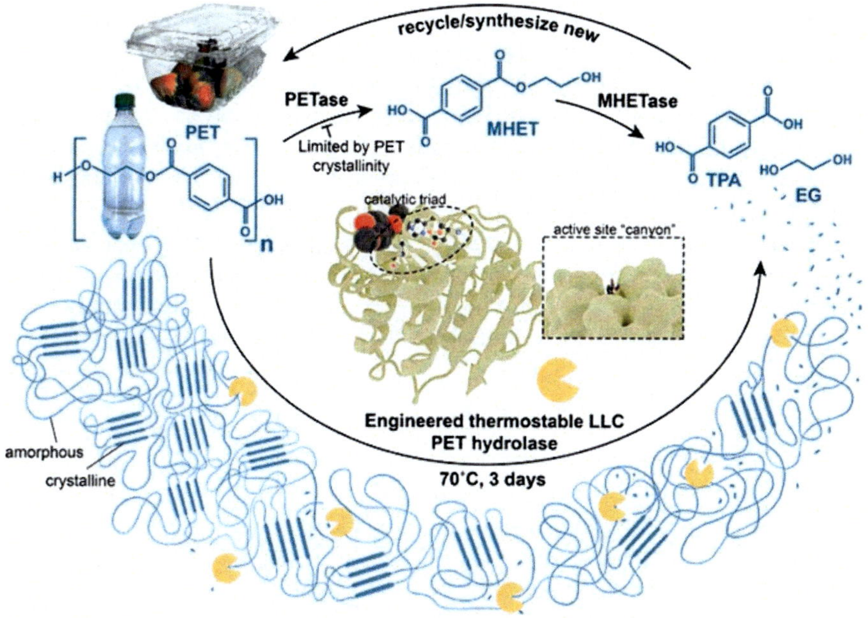

Fig. 6.4 Engineered PETases efficiently recycle PET to TPA. Enhanced serine hydrolases bind and degrade polymers with MHET and EG (monohydroxyethyl terephthalic acid, and EG ethylene glycol) [35]

methods by companies like Ioniqa, Jeplan, Teijin, Eastman, Ambercycle, Carbios, and Gr3n (microwaves). These techniques involve glycolysis, enzymatic hydrolysis, and microwave radiation to produce PET monomers (EG and TPA) or oligomers (BHET, DMT). Chemical recycling allows the production of virgin quality PET, but it is more costly and requires large-scale production for economic viability. Nevertheless, these companies are scaling up and validating their technologies to contribute to sustainable PET recycling solutions.

6.3.3 Polyurethane

Elastane, a polyurethane and polycondensation polymer extensively used in textiles, poses formidable recycling challenges within the textile industry. Presently, there are no well-established methods for recycling elastane on a pilot or demonstration scale. One of the primary reasons for this is the relatively minimal presence of elastane in (women's) fashion garments, which reduces the incentive for dedicated recycling initiatives. Consequently, current practices often do not involve recycling elastane in its original form. Nonetheless, researchers and industry experts are actively exploring potential solutions to enhance textile recycling, especially concerning the presence of elastane [38, 39]. One promising approach is solvolysis, which holds promise for polycondensation polymers like elastane. Solvolysis entails the use of solvents to break down the polymer, effectively separating elastane from other fibers present in the fabric. By removing the elastane component through solvolysis, the remaining fibers become more suitable for traditional recycling processes, thereby bolstering the potential for circularity in fashion production. Despite the inherent complexities and challenges surrounding elastane recycling, the fashion industry remains committed to finding sustainable and efficient solutions. As seen in Fig. 6.5, The polymers PIs showed exceptional mechanical properties, surpassing other polyimines reported in the literature (bio-PI3, bio-PI4, bio-PI5, and bio-PI6) and commonly used engineering plastics such as PLA, PET, PEF, PC, PU, PA6, PVC, PP, and HDPE.

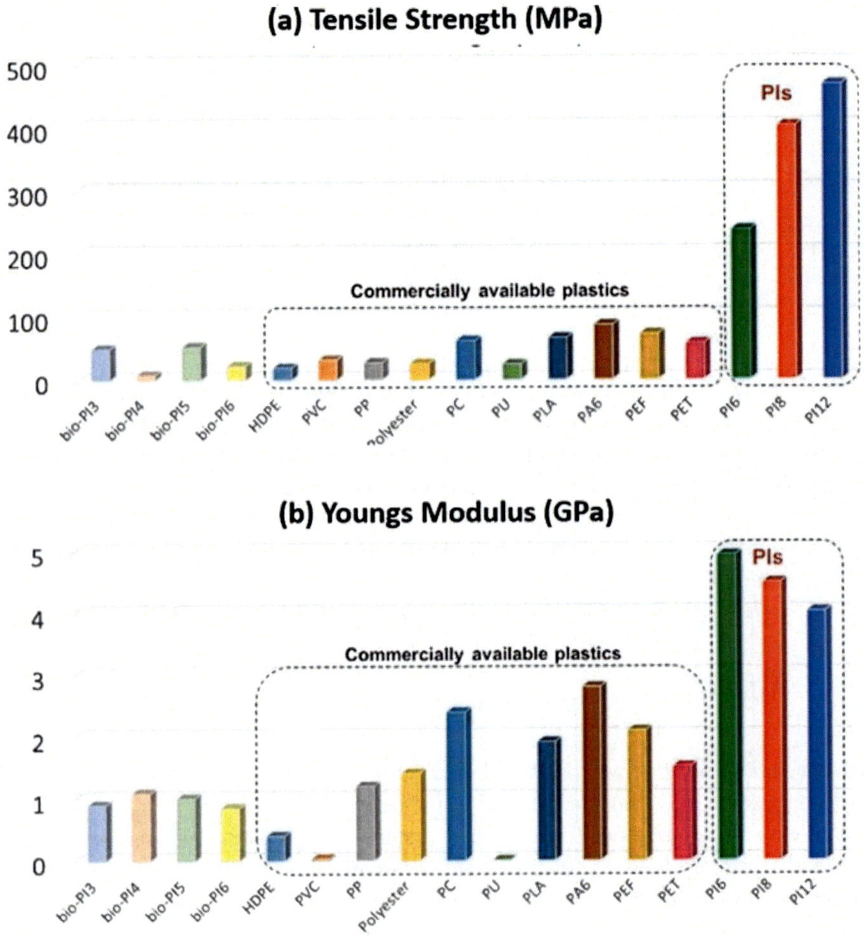

Fig. 6.5 Comparison of tensile strength and young's modulus of various polymers [40]

6.3.4 Polyolefins

Polyolefins, such as polyethylene (PE) or polypropylene (PP), are commonly used thermoplastics in various industries, including textiles. Although they have the potential for physical recycling through melting and re-spinning into new fibers and yarns, their application in apparel textiles remains limited. Chemical recycling methods, like depolymerization using solvolysis techniques, are not suitable for polyolefins due to the resilience of their chemical bonds. While polyolefins can be degraded at high temperatures through a free radical mechanism, this process does not generate re-usable monomers. Instead, it produces diverse mixtures of gases, liquids, and tar, which can be utilized as inputs for chemical cracking processes, contributing to the production of renewable bulk chemicals [41, 42]. Various companies and organizations, such as Veolia, Dow Chemical Company, and Borealis AG, have been actively working on recycling polyolefins and other plastics to support sustainability efforts. PlasticsEurope, the European Union, and the American Chemistry Council (ACC) are also prominent players promoting circular economy initiatives and plastic recycling, including polyolefins.

6.3.5 Polyamide

Polyamide fibers, found in both natural wool and synthetic nylon, present diverse recycling possibilities. Mechanical recycling effectively transforms wool's long fibers into new textiles through careful handling. On the other hand, nylon, a thermoplastic and polycondensation polymer, shares recycling potentials similar to polyesters. Successful mechanical recycling of post-consumer textiles is already demonstrated with wool, while nylon can be physically recycled by melting and reforming into fresh fibers. Fishing nets and nylon 6 carpets are essential inputs for physical recycling. Notably, fishing net recycling is technologically feasible, and the resulting recycled nylon 6 exhibits comparable properties to commercial nylons. Moreover, nylon 6 can be chemically recycled to recover its original building block, caprolactam, through an environmentally benign process [43, 44].

Companies involved in recycling polyamide fibers include Cardato, Boer Group, Geetanjali Woollens, and Novetex. For nylon recycling, Fulgar employs the MSC process to produce Q-Nova regenerated/virgin nylon 6,6 fiber, while Aquafil's Econyl technology depolymerizes nylon-6 fishing nets, carpets, and post-industrial textiles into caprolactam, producing Econyl yarn.

6.3.6 Polyacrylics

Polyacrylics, like wool, present opportunities for mechanical recycling, making it possible to recycle them in a manner comparable to wool fibers. The recycling process for polyacrylics includes essential steps such as colour sorting, cleaning, unravelling, and spinning the fibers to create new products. However, it is important to note that polyacrylics cannot be melted, which limits the application of certain recycling methods. As of now, there are no active efforts towards physical recycling methods for polyacrylics. This limitation arises from the fact that polyacrylics are formed through addition polymerization, preventing their depolymerization through solvolysis methods. Consequently, mechanical recycling remains the primary and practical approach for effectively recycling polyacrylic fibers, promoting sustainability and reducing waste in the textile industry [38].

Overall, natural polymers offer diverse recycling options for both fibers and polymers. Mechanical recycling suits staple fibers like cotton and wool. Physical recycling dissolves cellulose for new textile fibers. Condensation polymers like PET and nylon can be recycled through physical and chemical methods. Elastane has limited recycling options, and addition polymers, such as polyacrylics, have limited fiber recycling choices [38, 39].

6.4 Circular Models for Fashion Industries

These models revolve around closed-loop systems, where products are designed, manufactured, used, and eventually recycled or upcycled. Several frameworks support circularity in this industry, such as the rent, lease, or subscription model, allowing customers to rent clothing items for reuse, lessening overall consumption. Encouraging resale and second-hand markets enables customers to return used clothing, extending product lifecycles. The product-as-a-service (PaaS) approach offers fashion items as services with the company retaining ownership and responsibility for maintenance and recycling [45, 46].

Based on insights gathered from various literature sources, a notable trend is the adoption of circular economy models, where both recycled and virgin fibers are utilized. Fashion companies exert control over key aspects such as designing, packaging, distribution, and retail, while actively collaborating with NGOs and governmental agencies to achieve resource reduction goals. In the downstream phase, post-consumer garments are collected to initiate the recovery process. During this phase, fibers and materials undergo reprocessing and integration into new products [45, 47]. Traceability plays a pivotal role in ensuring transparency for both customers and recycling agents, allowing for an efficient circular flow. Additionally, sustainability reports serve as vital tools in enhancing transparency within the value chain. Overall, this emphasis on circularity and transparency signifies a growing commitment to sustainability in the fashion industry. In a recent study by Dragomir et al. the circular economy model (Fig. 6.6) for the fashion industry was framed, incorporating various circular solutions derived from the literature [47]. It distinguishes material and information flows within the value chain, aiming to answer two research questions: the practical solutions adopted by fashion retailers for circular business models and the comparison of companies in terms of circular economy implementation [48]. Prior literature has explored specific aspects of circularity, such as supplier compliance, sustainability attributes, and stakeholder engagement, but lacked a comprehensive framework [45].

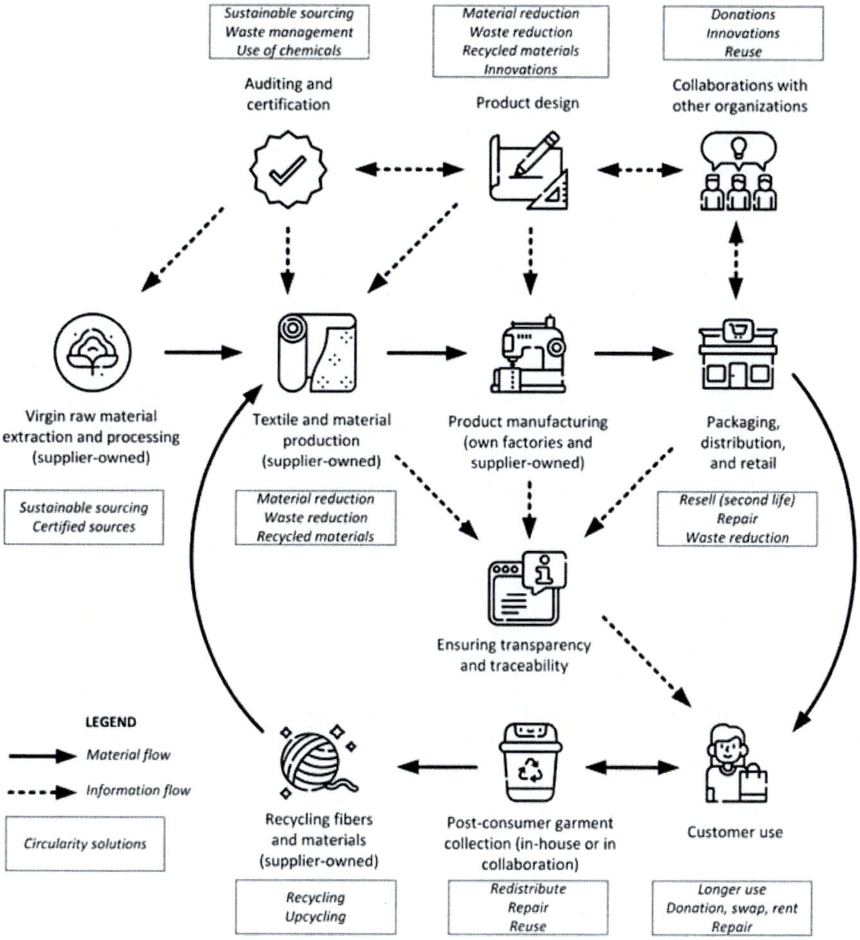

Fig. 6.6 The inputs, stages, and outputs of circularity in the fast fashion industry, highlighting the interconnected processes that promote sustainability, waste reduction, and resource efficiency throughout the value chain [45]

6.5 Challenges

Achieving circular fashion faces several challenges that hinder the industry's transition towards sustainability:

One major obstacle is the misalignment of metrics used to measure the success of circular programs. Brands often focus on sales volumes rather than the replacement rate, which is fundamental for circular business models. Without prioritizing the extent to which used or refurbished products can replace newly manufactured ones, the necessary shift towards reduced production volumes and circularity cannot be fully realized.

The fashion industry's historical emphasis on driving consumption through fast fashion and constant newness has ingrained a culture of relentless buying. Promoting sustainable practices solely to affluent consumers in niche product lines won't suffice to drive the mass market transformation needed for a truly circular fashion industry. Instead, concerted efforts should be made to recondition consumers to demand durable, sustainable apparel, fostering a mindset of reduced overall consumption [34].

Many clothing and footwear items are not designed with circularity in mind, lacking features that facilitate easy repair or recycling. Brands must adopt an eco-design approach, considering the environmental impacts of products across their entire life cycle. This means designing garments and footwear with repairability and recyclability in focus, ensuring they can seamlessly fit into circular business models.

Existing supply chains were purpose-built for linear production and distribution, posing challenges in achieving the necessary transparency and traceability for circularity. The complexity and global nature of these supply chains often hinder seamless integration into circular practices. To address this, brands must enhance transparency and traceability throughout the supply chain and build local networks that facilitate circular services and resource circulation.

Some of the other challenges in achieving circular fashion include the lack of efficient collection and sorting systems for used clothing, limited consumer awareness and education about circular practices, high costs of sustainable materials and technologies, long product lifecycles hindering demand for new circular products, limited access to post-consumer waste for recycling, regulatory and policy barriers, and a lack of collaboration and knowledge sharing within the industry. Addressing these challenges requires a comprehensive approach involving collaboration among all stakeholders in the fashion ecosystem to foster meaningful change and embrace circular design principles for a more sustainable and regenerative future [28].

6.6 Circular Approaches in Building Material Selection

Sustainable building material management involves a holistic approach from design to the building's entire life cycle. Designing for deconstruction facilitates future salvage of reusable materials during renovations or at end-of-life, reducing overall waste. Circular principles prioritize renewable, recycled, and recyclable materials, reducing environmental impact. Deconstructing existing buildings to salvage materials conserves resources and minimizes energy-intensive manufacturing. Efficient recycling systems process materials for reuse, like recycled concrete aggregates and glass cullet. Embracing circular approaches fosters a closed-loop system, minimizing waste and promoting sustainability in the construction industry [49]. As seen in Fig. 6.7, the global building material consumption tripled from 6.7 billion tons in 2000 to 17.5 billion tons in 2017 due to rapid urbanization and population growth. China experienced the most significant growth, accounting for over half of the global consumption in 2017, while Europe and North America stabilized or decreased their use during the same period.

Building materials have a complete life cycle from raw material extraction to circular processing through reuse, recycling, and recovery. From Fig. 6.8. The blue shading represents the life cycle of building materials, with the LCA boundary (dashed line) including input resources, building materials, and output emissions in different stages. Inventory data encompass energy and water usage, building materials, and polluting emissions. Details of LCA methodologies is covered in Chap. 2.

6.6.1 Concrete

Concrete recycling is a key aspect of circular approaches in the construction industry. When buildings or structures are demolished or renovated, a significant amount of concrete waste is generated. Instead of sending this waste to landfills, circular practices aim to recycle it to obtain recycled concrete aggregates (RCA). The recycling process begins with the collection of concrete debris, which is then transported to recycling facilities. At the recycling facility, the waste concrete is carefully sorted and cleaned to remove any contaminants or impurities. Next, the concrete is crushed into smaller pieces using specialized equipment such as jaw crushers or impact crushers. The crushed concrete is then screened to separate it into various sizes of aggregates. The resulting RCA has similar properties to natural aggregates and can be used as a substitute in various construction applications [52, 53]. By using RCA in new concrete mixtures, the demand for natural aggregates, such as gravel and sand, is reduced, leading to resource conservation, and minimizing the environmental impact of aggregate extraction.

Furthermore, recycling concrete waste into RCA helps in diverting substantial amounts of waste from landfills, contributing to waste reduction and more sustainable

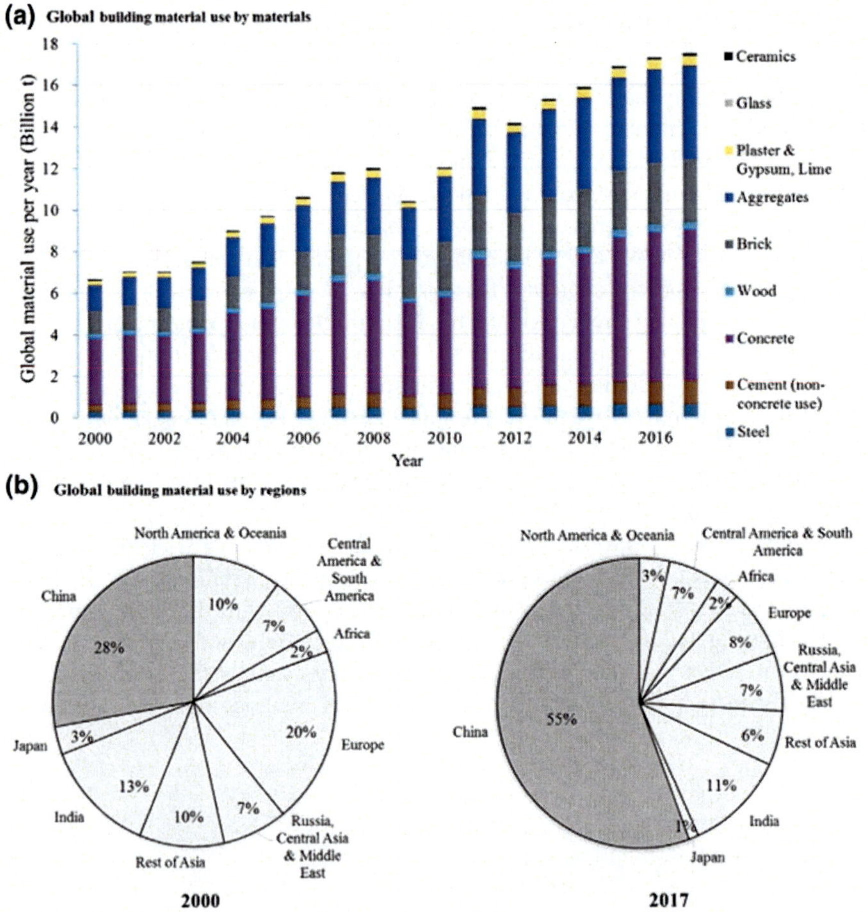

Fig. 6.7 Overview of global building material use (2000–2017) by material type and region. Analysis includes material consumption intensity country wise. [50]

waste management practices. The use of recycled concrete aggregates also results in lower greenhouse gas emissions compared to using virgin aggregates, making it an environmentally friendly choice (Table 6.4).

6.6.2 *Steel and Wood*

Steel is a highly recyclable material due to its inherent properties and widespread use in building construction. At the end of a building's life or during renovations, steel components can be carefully collected and processed for recycling. The recycling process involves the use of electric arc furnaces or oxygen furnaces to melt the

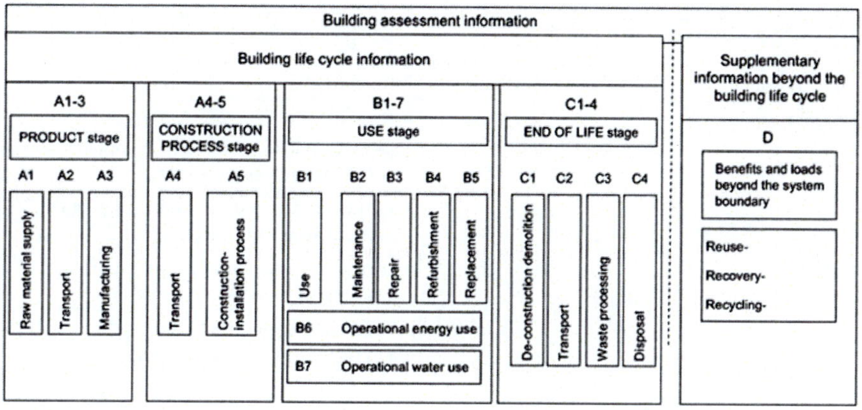

Fig. 6.8 Building Materials Life Cycle Assessment (LCA) Framework [51]

collected steel scrap. During the melting process, impurities and alloying elements are carefully controlled to produce high-quality recycled steel. The molten steel is then cast into various forms, such as beams, columns, and other structural elements, for use in new construction projects. The recycling process can be repeated multiple times without compromising the strength and properties of the steel, making it an infinitely recyclable material. One of the key benefits of steel recycling is its significant environmental impact reduction. By recycling steel, the need for mining and extracting raw iron ore, a resource-intensive process that contributes to greenhouse gas emissions and habitat destruction, is greatly minimized. Additionally, recycling steel consumes less energy compared to producing steel from raw materials, leading to reduced carbon emissions and conserving energy resources [53, 54].

When choosing wood materials, opting for certified timber from responsibly managed forests ensures that wood resources are harvested in an environmentally responsible and sustainable manner. Certification schemes like the Forest Stewardship Council (FSC) and the Programme for the Endorsement of Forest Certification (PEFC) provide assurance that the wood comes from well-managed forests, where trees are harvested in a manner that promotes forest regeneration and biodiversity conservation. Engineered wood products, such as plywood and oriented strand board (OSB), play a crucial role in optimizing wood resources' efficient use. These products are manufactured by bonding wood strands or veneers with adhesives, creating strong and versatile materials suitable for various construction applications [55, 56]. Unlike solid wood, engineered wood products utilize smaller wood pieces effectively, reducing waste and maximizing the use of available wood resources. The production of engineered wood products also requires less energy compared to traditional wood processing methods, contributing to a more sustainable and resource-efficient building industry.

Table 6.4 List of companies and their initiatives for improving the sustainability of building material

Company Name	Country	Initiatives
LafargeHolcim	Global	• Establishing recycling facilities to produce recycled concrete aggregates (RCA) from waste concrete
		• Reducing demand for natural aggregates and diverting waste from landfills
CEMEX	Global	• Investing in advanced technologies and equipment for concrete recycling
		• Producing high-quality recycled aggregates to support sustainable construction
Recycled Aggregate Products (RAP)	Australia	• Operating recycling facilities to convert demolished concrete into RCA
		• Collaborating with construction companies for responsible concrete waste management
Delta Group	Australia	• Prioritizing sustainable construction practices and operating concrete recycling facilities
		• Recycling significant volumes of concrete waste to reduce environmental impacts
Hanson	Global	• Incorporating recycled concrete aggregates in their concrete mixes
		• Demonstrating commitment to sustainability and environmental responsibility
Sika	Global	• Offering concrete admixtures and construction solutions to enhance concrete durability and performance
		• Supporting longer service life and reduced material consumption

6.6.3 Other Materials

Glass, being infinitely recyclable, allows for a closed-loop system where recycled glass (cullet) is blended with new raw materials to create new glass products, reducing energy consumption and conserving resources. Circular practices extend to bricks and masonry, with salvaged materials from deconstructed buildings reintegrated into new construction, minimizing the need for new bricks and reducing environmental impact. Emphasizing the use of recycled content, circular approaches apply to insulation materials, such as recycled fiberglass and cellulose insulation, reducing waste and promoting energy efficiency. In the realm of plastics, circular building practices prioritize reducing single-use plastics and incorporating recycled plastic in construction materials like decking and insulation. Lastly, circular strategies extend to roofing materials, encouraging the use of recycled asphalt shingles and metal roofing to contribute to a more sustainable construction industry [57].

6.7 Critical Parameters and KPIs in Building Materials

Critical parameters in building materials include thermal conductivity, thermal diffusivity, specific heat capacity, porosity, permeability, vapor permeability (WVTR), sorption isotherms, water absorption, capillary action, hygroscopicity, water vapor sorption hysteresis, vapor–liquid interaction, diffusion coefficients, adsorption and absorption, mass permeability, air permeability, and air resistance. These properties impact insulation, energy efficiency, moisture management, durability, and indoor air quality in buildings. Thermal Conductivity, Thermal Diffusivity, and Specific Heat Capacity affect how materials conduct and store heat, impacting insulation, energy efficiency, and thermal responsiveness. Porosity and Permeability are linked to moisture absorption and transport, influencing moisture management and durability. Vapor Permeability and Sorption Isotherms are crucial for managing moisture in buildings, while Water Absorption and Capillary Action relate to water resistance and potential damage. Hygroscopicity and Water Vapor Sorption Hysteresis are related to moisture content and fluctuations, affecting humidity control. Vapor–Liquid Interaction and Diffusion Coefficients impact moisture transport and vapor control. Adsorption and Absorption influence moisture management, and Mass Permeability and Air Permeability are related to air quality and ventilation. Air Resistance impacts building ventilation and thermal comfort. Understanding and optimizing these parameters are vital for ensuring efficient and sustainable construction materials and building performance [58].

Porosity is a vital factor in building materials' durability and sustainability, defined as the ratio of pores' area or volume to the total area or volume of the material. It can be absolute, considering all pores, or effective, focusing on connected pores. Various investigative techniques characterize porosity, including capillary and gel C–S–H porosity in cement paste. Aggregates in mortar and concrete also influence porosity near grain surfaces. Properly understanding and managing porosity is essential for ensuring sustainable and long-lasting building materials. A crucial aspect of ensuring the sustainability of civil-engineering structures lies in understanding diffusion properties. Diffusion involves the movement of components in a medium, driven by concentration differences from high to low areas, even in stationary fluids. This process is irreversible and challenging to track at the molecular level, requiring macroscopic descriptions. The first Fick law quantifies mass transport per unit area and time, with the diffusion coefficient (D) characterizing species diffusivity in the medium. Effective diffusivity (D_e) considers porosity in heterogeneous media, influencing ion flow. High-porosity materials exhibit lower De compared to clear fluid diffusivity (D), and mineral additives can significantly reduce diffusivity [59, 60]. Understanding material diffusivity is vital for ensuring their durability and sustainability in construction endeavours.

Additionally, porous media possess interconnected properties of porosity and permeability. Permeability gauges fluid flow in a medium under pressure gradients, while effective porosity reflects hydraulic permeability, influenced by pore structure, interconnectivity, and size distribution. Porous materials exhibit anisotropy and heterogeneity due to varying properties. Cracks act as preferential flow paths, affecting diffusion and permeability, vital for material durability and structural performance. Analyzing permeability's impact on cracked porous volumes unveils changes in pressure gradients and flow, crucial for material behavior. Understanding these complexities ensures construction materials' sustainability and safety.

Further, the amplification of permeability by cracks and the water–vapor transfer and adsorption phenomena has significant implications for water storage and diffusion. These interactions with the surrounding environment give materials moisture-buffering capacity and the capability to moderate relative humidity variations. The inclusion of biomaterials adds complexity to the interaction on multiple scales and alters the material's durability. The inhomogeneity between biomaterials and cement paste affects durability and humidity distribution in innovative construction materials, contributing to the development of fungi and its potential impact on human health [61].

LCA aims to assess the environmental burdens and resource consumption at each stage, from the extraction of raw materials, manufacturing, transportation, construction, and use, to waste management and disposal. By analyzing these stages, LCA helps identify the key environmental impacts and potential mitigation strategies (Fig. 6.9) to reduce the overall ecological footprint of building materials. The first principle of LCA is to comprehensively assess the impacts of each life cycle process activity, including energy and resource consumption, as well as pollutant emissions. In the context of building materials, it is crucial to understand the environmental implications of different activities throughout their life cycle to make informed decisions that prioritize sustainability and resource efficiency. The Key Performance Indicators (KPIs) in LCA of building materials include Global Warming Potential (GWP), Energy Use, Acidification Potential, Eutrophication Potential, Ozone Depletion Potential, Resource Depletion, Water Footprint, Land Use, Human Health Damage, Biodiversity Impact, and Waste Generation.

Notably, the extraction and manufacturing of building materials are major contributors to environmental pollutants, accounting for nearly 90% of the total life cycle impacts. Activities such as mining, quarrying, and processing of raw materials can lead to significant emissions of greenhouse gases, particulate matter, and other pollutants that affect air and water quality. Additionally, transportation and construction stages also contribute to environmental impacts, particularly through emissions of nitrogen oxides (NOx) and carbon dioxide (CO_2) resulting from the use of fossil fuels in vehicles and machinery. These emissions contribute to air pollution and climate change. Waste treatment, another critical stage in the life cycle of building materials, poses its own environmental challenges. Waste plaster and wood, commonly generated during construction and demolition activities, can release organic acids in landfills. Incineration of wood, plastic, and paper waste can lead to the generation of

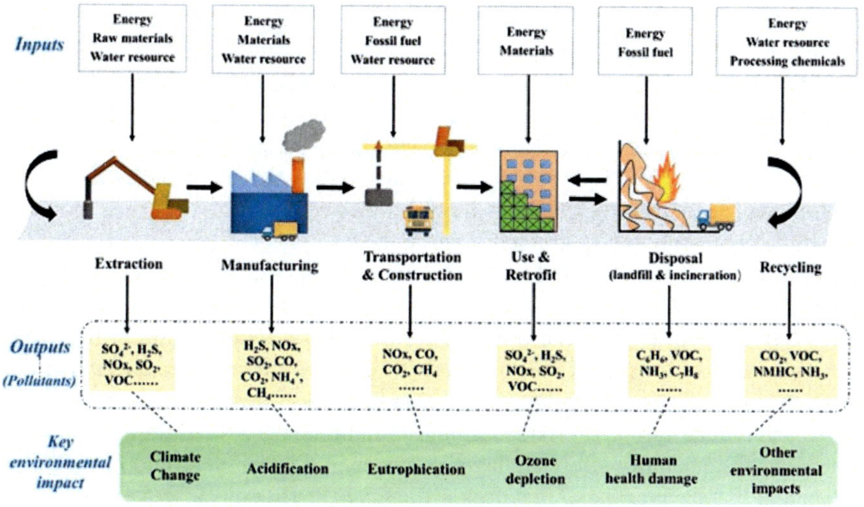

Fig. 6.9 Environmental Impacts of Building Materials' Life Cycle [50]

pollutants such as ammonia (NH_3), heavy metal ions, and volatile organic compounds (VOCs), which have adverse effects on human health and the environment.

6.8 Global Initiatives for Sustainable Building Materials and Fashion

Global initiatives for building materials and fashion have been gaining momentum to address sustainability and environmental concerns. Initiatives such as green building certifications like LEED (Leadership in Energy and Environmental Design) and BREEAM (Building Research Establishment Environmental Assessment Method), Digital Product Passport (DPP) [62] have been widely adopted to encourage the use of renewable materials, energy-efficient designs, and environmentally conscious construction practices. Additionally, there is a push for the development of low-carbon and recycled building materials, such as eco-friendly concrete alternatives, sustainable timber, and innovative insulation materials, to reduce the environmental impact of construction activities. These initiatives also prioritize the use of local materials and sourcing practices to minimize transportation-related carbon emissions.

In the fashion industry, sustainable initiatives aim to address the environmental and ethical challenges associated with the production and consumption of clothing. Fast fashion, characterized by its rapid production and disposal of garments, has led to significant environmental impacts, including excessive water usage, pollution, and waste generation. To combat these issues, numerous initiatives have emerged to promote sustainable fashion practices. These include the use of eco-friendly and organic fabrics, adopting circular economy principles to promote recycling and upcycling of clothing, and encouraging ethical labor practices throughout the supply chain. Sustainable fashion initiatives also emphasize the importance of consumer awareness and education to promote responsible purchasing behavior and reduce overconsumption. Additionally, several global fashion brands have committed to sustainability goals, aiming to reduce their carbon footprint, eliminate harmful chemicals from production processes, and ensure fair labor practices [28].

Interface Inc., Saint-Gobain, LafargeHolcim, and Skanska are prominent players in the building materials sector committed to sustainability. Interface focuses on recycled and bio-based materials, aiming for a carbon-negative footprint by 2040. Saint-Gobain prioritizes renewable materials and energy-efficient solutions to reduce greenhouse gas emissions. LafargeHolcim aims to develop low-carbon concrete alternatives and reduce carbon emissions. In the fashion industry, several global brands are leading sustainability efforts. Patagonia pioneers' sustainable fashion with recycled materials and ethical labor practices. EILEEN FISHER emphasizes eco-friendly fabrics and circular economy principles. H&M's Conscious Collection offers sustainable clothing made from organic cotton and recycled polyester. Stella McCartney avoids fur and leather, promoting ethical practices. Levi's adopts sustainable denim production and encourages recycling and upcycling of jeans.

6.9 Conclusion

The construction industry continuously seeks sustainable improvements through extensive research and development. Innovations in materials and techniques have led to impressive and environmentally friendly buildings. However, the journey from R&D to practical implementation takes time. Adopting a performance-based approach instead of a traditional one, a life-cycle assessment helps identify the most polluting stages of concrete's life, paving the way for targeted improvements. Concrete's thermal mass proves beneficial during its operational phase and demolition, compensating for its initial unsustainability during the construction phase. Reducing CO_2 emissions from building materials production is vital, primarily arising from clinker heating. Various options, such as clinker replacement, alternative fuels, improved kilns, carbon capture, and recycled materials, can be explored to mitigate these emissions. The use of supplementary cementitious materials (SCMs) is a simple technique to reduce CO_2 emissions and has already gained traction in the building materials industry. Research should focus on enhancing their efficiency, although early strength issues need to be addressed. Incorporating recycled and secondary aggregates can further reduce the carbon footprint and conserve natural resources. Additionally, understanding materials CO2 absorption and its impact on recycled aggregate quality requires further investigation.

When comparing different building materials, their moisture management capabilities are of paramount importance. Materials with high vapor permeability, such as breathable membranes and permeable concrete, can effectively manage moisture within the building envelope, preventing potential issues like mold growth and degradation. Moreover, the consideration of diffusion coefficients allows for an understanding of how gases and moisture migrate through materials, influencing indoor air quality and the overall building performance.

In the fashion realm, fabrics with a larger surface area, such as lightweight and porous materials, generally require fewer raw materials in their production, resulting in reduced resource consumption and waste generation. Breathability is another essential parameter in textile fashion that influences comfort and wearer satisfaction. Fabrics with high breathability, like natural fibers such as cotton and linen, allow for better air circulation, reducing the likelihood of moisture buildup and odors. This not only enhances the comfort of the wearer but also promotes a longer product lifespan. Volume considerations are crucial in the context of circularity, as bulky and voluminous textiles may require more space and energy during transportation and storage.

Thus, the adoption of eco-friendly materials and circular business models will not only contribute to a greener fashion and construction industry but also create a positive impact on the planet and society.

Activity: MCQs

1. What is the primary goal of circular approaches in fashion industries and building materials?
   (a) To increase waste generation
   (b) To promote single-use materials
   (c) To reduce resource consumption and waste
   (d) To encourage landfilling of textiles and construction materials
2. Which of the following is NOT a circular practice in the fashion industry?
   (a) Fast fashion production
   (b) Recycling textile fibers for new garments
   (c) Renting or leasing clothing items
   (d) Upcycling old clothing into new designs
3. How does circularity in building materials contribute to sustainability?
   (a) By increasing resource extraction
   (b) By promoting short-term use of construction materials
   (c) By reducing the need for new raw materials
   (d) By encouraging waste dumping in landfills
4. Which circular approach can be considered for sustainable fashion consumption?
   (a) Frequent disposal of clothing items
   (b) Avoiding repair and reuse of garments
   (c) Extending the lifespan of clothing through maintenance

   (d) Buying new clothing items frequently

5. Which of the following is an example of circularity in the fashion industry?
   (a) Donating used clothing to landfills
   (b) Incinerating old garments
   (c) Reusing fabrics from old clothes to make new garments
   (d) Using single-use plastics for clothing packaging
6. What is a key benefit of adopting circular practices in the fashion industry and building materials?
   (a) Increased resource consumption
   (b) Higher waste generation
   (c) Greater cost savings and sustainability
   (d) Limited product choices for consumers
7. Which circular approach in building materials involves using recycled materials in construction projects?
   (a) Landfilling of construction waste
   (b) Incineration of old building materials
   (c) Reusing and repurposing construction waste
   (d) Rapid disposal of building materials
8. How does circularity in fashion industries contribute to waste reduction?
   (a) By promoting fast fashion trends
   (b) By encouraging single-use clothing
   (c) By encouraging clothing disposal after a few wears
   (d) By promoting clothing repair, resale, and recycling
9. Which of the following is a circular practice in the fashion industry that prioritizes sustainability?
   (a) Fast fashion production with short product lifespans
   (b) Using non-recyclable and non-renewable materials
   (c) Adopting eco-friendly and biodegradable fabrics
   (d) Ignoring the environmental impact of garment production
10. Circular approaches in fashion industries and building materials aim to achieve:
   (a) Increased waste generation
   (b) Greater resource consumption
   (c) Sustainable and efficient resource utilization
   (d) Escalating waste dumping in landfills

Answers:

1. (c) To reduce resource consumption and waste
2. (a) Fast fashion production
3. (a) By reducing the need for new raw materials

4. (a) Extending the lifespan of clothing through maintenance
5. (a) Reusing fabrics from old clothes to make new garments
6. (c) Greater cost savings and sustainability
7. (c) Reusing and repurposing construction waste
8. (d) By promoting clothing repair, resale, and recycling
9. (c) Adopting eco-friendly and biodegradable fabrics
10. (c) Sustainable and efficient resource utilization

References

1. R. Aus, H. Moora, M. Vihma, R. Unt, M. Kiisa, S. Kapur, Designing for circular fashion: integrating upcycling into conventional garment manufacturing processes, Fashion and Textiles. 8 (2021) 1–18. https://doi.org/https://doi.org/10.1186/S40691-021-00262-9/FIGURES/7.
2. E. D'Itria, R. Aus, Circular fashion: evolving practices in a changing industry, Https://Doi.Org/https://doi.org/10.1080/15487733.2023.2220592. 19 (2023) 2220592. https://doi.org/10.1080/15487733.2023.2220592.
3. M.M. Wagner, T. Heinzel, Human Perceptions of Recycled Textiles and Circular Fashion: A Systematic Literature Review, Sustainability 2020, Vol. 12, Page 10599. 12 (2020) 10599. https://doi.org/10.3390/SU122410599.
4. E. Hultberg, R. Pal, Lessons on business model scalability for circular economy in the fashion retail value chain: Towards a conceptual model, Sustain Prod Consum. 28 (2021) 686–698. https://doi.org/https://doi.org/10.1016/J.SPC.2021.06.033.
5. The Circular Fashion Economy: A Sustainable Shift Towards a Greener Future - EcoMatcher, (n.d.). https://www.ecomatcher.com/the-circular-fashion-economy-a-sustainable-shift-towards-a-greener-future/ (accessed July 20, 2023).
6. J.P. Juanga-Labayen, I. V. Labayen, Q. Yuan, A Review on Textile Recycling Practices and Challenges, Textiles 2022, Vol. 2, Pages 174–188. 2 (2022) 174–188. https://doi.org/10.3390/TEXTILES2010010.
7. A. De Aguiar Hugo, J. De Nadae, R. Da, S. Lima, Y. Shimomura, S. Hosono, Can Fashion Be Circular? A Literature Review on Circular Economy Barriers, Drivers, and Practices in the Fashion Industry's Productive Chain, Sustainability 2021, Vol. 13, Page 12246. 13 (2021) 12246. https://doi.org/10.3390/SU132112246.
8. C.W. Ki, S.M. Chong, J.E. Ha-Brookshire, How fashion can achieve sustainable development through a circular economy and stakeholder engagement: A systematic literature review, Corp Soc Responsib Environ Manag. 27 (2020) 2401–2424. https://doi.org/https://doi.org/10.1002/CSR.1970.
9. A. Appolloni, G. Centi, N. Yang, Promoting carbon circularity for a sustainable and resilience fashion industry, Curr Opin Green Sustain Chem. 39 (2023) 100719. https://doi.org/https://doi.org/10.1016/J.COGSC.2022.100719.
10. P. Harmsen, M. Scheffer, H. Bos, Textiles for Circular Fashion: The Logic behind Recycling Options, Sustainability 2021, Vol. 13, Page 9714. 13 (2021) 9714. https://doi.org/10.3390/SU13179714.
11. T. Brydges, Closing the loop on take, make, waste: Investigating circular economy practices in the Swedish fashion industry, J Clean Prod. 293 (2021) 126245. https://doi.org/https://doi.org/10.1016/J.JCLEPRO.2021.126245.
12. F. Welle, Investigation into cross-contamination during cleaning efficiency testing in PET recycling, Resour Conserv Recycl. 112 (2016) 65–72. https://doi.org/https://doi.org/10.1016/J.RESCONREC.2016.05.003.

13. E. Barnard, J.J. Rubio Arias, W. Thielemans, Chemolytic depolymerisation of PET: a review, Green Chemistry. 23 (2021) 3765–3789. https://doi.org/10.1039/D1GC00887K.
14. R.L. Smith, S. Takkellapati, R.C. Riegerix, Recycling of Plastics in the United States: Plastic Material Flows and Polyethylene Terephthalate (PET) Recycling Processes, ACS Sustain Chem Eng. 10 (2022) 2084–2096. https://doi.org/https://doi.org/10.1021/ACSSUSCHEMENG.1C06845.
15. L. Luo, Y. Zhou, X. Xu, W. Shi, J. Hu, G. Li, X. Qu, Y. Guo, X. Tian, A. Zaman, D. Hui, Z. Zhou, Progress in construction of bio-inspired physico-antimicrobial surfaces, Nanotechnol Rev. 9 (2020) 1562–1575. https://doi.org/https://doi.org/10.1515/NTREV-2020-0089.
16. I.K. Voets, From ice-binding proteins to bio-inspired antifreeze materials, Soft Matter. 13 (2017) 4808–4823. https://doi.org/https://doi.org/10.1039/C6SM02867E.
17. R.M. Hegde, R.M. Rego, K.M. Potla, M.D. Kurkuri, M. Kigga, Bio-inspired materials for defluoridation of water: A review, Chemosphere. 253 (2020) 126657. https://doi.org/https://doi.org/10.1016/J.CHEMOSPHERE.2020.126657.
18. Y. Shi, J. Zhang, L. Pan, Y. Shi, G. Yu, Energy gels: A bio-inspired material platform for advanced energy applications, Nano Today. 11 (2016) 738–762. https://doi.org/https://doi.org/10.1016/J.NANTOD.2016.10.002.
19. C.H. Jo, N. Voronina, Y.K. Sun, S.T. Myung, Gifts from Nature: Bio-Inspired Materials for Rechargeable Secondary Batteries, Advanced Materials. 33 (2021) 2006019. https://doi.org/https://doi.org/10.1002/ADMA.202006019.
20. F. Buccino, I. Aiazzi, A. Casto, B. Liu, M.C. Sbarra, G. Ziarelli, L.M. Vergani, S. Bagherifard, Down to the Bone: A Novel Bio-Inspired Design Concept, Materials 2021, Vol. 14, Page 4226. 14 (2021) 4226. https://doi.org/10.3390/MA14154226.
21. Y. Geng, K. Jiao, X. Liu, P. Ying, O. Odunmbaku, Y. Zhang, S.C. Tan, L. Li, W. Zhang, M. Li, Applications of bio-derived/bio-inspired materials in the field of interfacial solar steam generation, Nano Research 2021 15:4. 15 (2021) 3122–3142. https://doi.org/10.1007/S12274-021-3834-9.
22. N.K. Katiyar, G. Goel, S. Hawi, S. Goel, Nature-inspired materials: Emerging trends and prospects, NPG Asia Materials 2021 13:1. 13 (2021) 1–16. https://doi.org/10.1038/s41427-021-00322-y.
23. N. ElShishtawy, P. Sinha, J.A. Bennell, A comparative review of zero-waste fashion design thinking and operational research on cutting and packing optimisation, Https://Doi.Org/https://doi.org/10.1080/17543266.2021.1990416. 15 (2021) 187–199. https://doi.org/10.1080/17543266.2021.1990416.
24. H. McQuillan, Digital 3D design as a tool for augmenting zero-waste fashion design practice, Https://Doi.Org/https://doi.org/10.1080/17543266.2020.1737248. 13 (2020) 89–100. https://doi.org/10.1080/17543266.2020.1737248.
25. E. Saeidi, V.S. Wimberley, Precious cut: exploring creative pattern cutting and draping for zero-waste design, Https://Doi.Org/https://doi.org/10.1080/17543266.2017.1389997. 11 (2017) 243–253. https://doi.org/10.1080/17543266.2017.1389997.
26. R. Rahman, S.Z.F.S. Putra, Tensile properties of natural and synthetic fiber-reinforced polymer composites, Mechanical and Physical Testing of Biocomposites, Fibre-Reinforced Composites and Hybrid Composites. (2019) 81–102. https://doi.org/10.1016/B978-0-08-102292-4.00005-9.
27. Q. Cheng, Y. Guo, Z. Liu, G. Zhang, P. Gu, A new modularization method of heavy-duty machine tool for green remanufacturing, Https://Doi.Org/https://doi.org/10.1177/0954406217752025. 232 (2018) 4237–4254. https://doi.org/10.1177/0954406217752025.
28. L. Lindkvist Haziri, E. Sundin, Supporting design for remanufacturing - A framework for implementing information feedback from remanufacturing to product design, Journal of Remanufacturing. 10 (2020) 57–76. https://doi.org/10.1007/S13243-019-00074-7/TABLES/5.
29. J.A. Fadeyi, L. Monplaisir, C. Aguwa, The integration of core cleaning and product serviceability into product modularization for the creation of an improved remanufacturing-product service system, J Clean Prod. 159 (2017) 446–455. https://doi.org/https://doi.org/10.1016/J.JCLEPRO.2017.05.083.

30. H. Dahlbo, K. Aalto, H. Eskelinen, H. Salmenperä, Increasing textile circulation—Consequences and requirements, Sustain Prod Consum. 9 (2017) 44–57. https://doi.org/https://doi.org/10.1016/J.SPC.2016.06.005.
31. R.P. Singh, P.K. Singh, R. Gupta, R.L. Singh, Treatment and Recycling of Wastewater from Textile Industry, (2019) 225–266. https://doi.org/10.1007/978-981-13-1468-1_8.
32. Y.J. Cai, Y. Chen, T. Siqin, T.M. Choi, S.H. Chung, Pay upfront or pay later? Fixed royal payment in sustainable fashion brand franchising, Int J Prod Econ. 214 (2019) 95–105. https://doi.org/https://doi.org/10.1016/J.IJPE.2019.03.025.
33. O.A. El Seoud, M. Kostag, K. Jedvert, N.I. Malek, Cellulose Regeneration and Chemical Recycling: Closing the "Cellulose Gap" Using Environmentally Benign Solvents, Macromol Mater Eng. 305 (2020) 1900832. https://doi.org/https://doi.org/10.1002/MAME.201900832.
34. J.M. Rieland, B.J. Love, Ionic liquids: A milestone on the pathway to greener recycling of cellulose from biomass, Resour Conserv Recycl. 155 (2020) 104678. https://doi.org/https://doi.org/10.1016/J.RESCONREC.2019.104678.
35. M. O-Reilly, J.A. Stubbe, PET polymer recycling, Biochemistry. 59 (2020) 2316–2318. https://doi.org/10.1021/ACS.BIOCHEM.0C00457/ASSET/IMAGES/LARGE/BI0C00457_0002.JPEG.
36. J. Payne, M.D. Jones, The Chemical Recycling of Polyesters for a Circular Plastics Economy: Challenges and Emerging Opportunities, ChemSusChem. 14 (2021) 4041–4070. https://doi.org/https://doi.org/10.1002/CSSC.202100400.
37. C. Pudack, M. Stepanski, P. Fässler, PET Recycling – Contributions of Crystallization to Sustainability, Chemie Ingenieur Technik. 92 (2020) 452–458. https://doi.org/https://doi.org/10.1002/CITE.201900085.
38. M. Chanda, Chemical aspects of polymer recycling, Advanced Industrial and Engineering Polymer Research. 4 (2021) 133–150. https://doi.org/https://doi.org/10.1016/J.AIEPR.2021.06.002.
39. O. Valerio, R. Muthuraj, A. Codou, Strategies for polymer to polymer recycling from waste: Current trends and opportunities for improving the circular economy of polymers in South America, Curr Opin Green Sustain Chem. 25 (2020) 100381. https://doi.org/https://doi.org/10.1016/J.COGSC.2020.100381.
40. N. Kasmi, E. Bäckström, M. Hakkarainen, Open-loop recycling of post-consumer PET to closed-loop chemically recyclable high-performance polyimines, Resour Conserv Recycl. 193 (2023) 106974. https://doi.org/https://doi.org/10.1016/J.RESCONREC.2023.106974.
41. H. Huang, W. Liu, Z. Liu, An additive manufacturing-based approach for carbon fiber reinforced polymer recycling, CIRP Annals. 69 (2020) 33–36. https://doi.org/https://doi.org/10.1016/J.CIRP.2020.04.085.
42. V. Shanmugam, O. Das, R.E. Neisiany, K. Babu, S. Singh, M.S. Hedenqvist, F. Berto, S. Ramakrishna, Polymer Recycling in Additive Manufacturing: an Opportunity for the Circular Economy, Materials Circular Economy 2020 2:1. 2 (2020) 1–11. https://doi.org/10.1007/S42824-020-00012-0.
43. J. Datta, K. Błażek, M. Włoch, R. Bukowski, A New Approach to Chemical Recycling of Polyamide 6.6 and Synthesis of Polyurethanes with Recovered Intermediates, J Polym Environ. 26 (2018) 4415–4429. https://doi.org/10.1007/S10924-018-1314-4/TABLES/6.
44. R.J. Tapper, M.L. Longana, I. Hamerton, K.D. Potter, A closed-loop recycling process for discontinuous carbon fibre polyamide 6 composites, Compos B Eng. 179 (2019) 107418. https://doi.org/https://doi.org/10.1016/J.COMPOSITESB.2019.107418.
45. V.D. Dragomir, M. Dumitru, Practical solutions for circular business models in the fashion industry, Cleaner Logistics and Supply Chain. 4 (2022) 100040. https://doi.org/https://doi.org/10.1016/J.CLSCN.2022.100040.
46. P.H. Huynh, "Enabling circular business models in the fashion industry: the role of digital innovation," International Journal of Productivity and Performance Management. 71 (2022) 870–895. https://doi.org/https://doi.org/10.1108/IJPPM-12-2020-0683/FULL/XML.
47. A. Abdelmeguid, M. Afy-Shararah, K. Salonitis, Investigating the challenges of applying the principles of the circular economy in the fashion industry: A systematic review, Sustain Prod Consum. 32 (2022) 505–518. https://doi.org/https://doi.org/10.1016/J.SPC.2022.05.009.

48. M. Colucci, A. Vecchi, Close the loop: Evidence on the implementation of the circular economy from the Italian fashion industry, Bus Strategy Environ. 30 (2021) 856–873. https://doi.org/https://doi.org/10.1002/BSE.2658.
49. K.M. Rahla, R. Mateus, L. Bragança, Implementing Circular Economy Strategies in Buildings—From Theory to Practice, Applied System Innovation 2021, Vol. 4, Page 26. 4 (2021) 26. https://doi.org/10.3390/ASI4020026.
50. B. Huang, X. Gao, X. Xu, J. Song, Y. Geng, J. Sarkis, T. Fishman, H. Kua, J. Nakatani, A Life Cycle Thinking Framework to Mitigate the Environmental Impact of Building Materials, One Earth. 3 (2020) 564–573. https://doi.org/https://doi.org/10.1016/J.ONEEAR.2020.10.010/ATTACHMENT/21466AA9-D9C5-4970-8BCF-99FCF0DB5908/MMC1.PDF.
51. T. Potrč Obrecht, R. Kunič, S. Jordan, A. Legat, Roles of the reference service life (RSL) of buildings and the RSL of building components in the environmental impacts of buildings, IOP Conf Ser Earth Environ Sci. 323 (2019) 012146. https://doi.org/10.1088/1755-1315/323/1/012146.
52. L.C.M. Eberhardt, M. Birkved, H. Birgisdottir, Building design and construction strategies for a circular economy, Architectural Engineering and Design Management. 18 (2022) 93–113. https://doi.org/https://doi.org/10.1080/17452007.2020.1781588.
53. L.A. Akanbi, L.O. Oyedele, O.O. Akinade, A.O. Ajayi, M. Davila Delgado, M. Bilal, S.A. Bello, Salvaging building materials in a circular economy: A BIM-based whole-life performance estimator, Resour Conserv Recycl. 129 (2018) 175–186. https://doi.org/10.1016/J.RESCONREC.2017.10.026.
54. A. Soni, P.K. Das, S. Kumar, Selection of waste plastics for incorporation in agro-waste as sustainable building construction materials concerning circular economy: an integrated MCDM approach, Clean Technol Environ Policy. (2023) 1–21. https://doi.org/10.1007/S10098-023-02539-7/METRICS.
55. L. Akanbi, L. Oyedele, J.M.D. Delgado, M. Bilal, O. Akinade, A. Ajayi, N. Mohammed-Yakub, Reusability analytics tool for end-of-life assessment of building materials in a circular economy, World Journal of Science, Technology and Sustainable Development. 16 (2019) 40–55. https://doi.org/https://doi.org/10.1108/WJSTSD-05-2018-0041/FULL/XML.
56. M. Zairul, The recent trends on prefabricated buildings with circular economy (CE) approach, Clean Eng Technol. 4 (2021) 100239. https://doi.org/https://doi.org/10.1016/J.CLET.2021.100239.
57. C. Gillott, W. Mihkelson, M. Lanau, D. Cheshire, D. Densley Tingley, Developing Regenerate: A circular economy engagement tool for the assessment of new and existing buildings, J Ind Ecol. 27 (2023) 423–435. https://doi.org/10.1111/JIEC.13377.
58. N. Ahn, A. Dodoo, M. Riggio, L. Muszynski, L. Schimleck, M. Puettmann, Circular economy in mass timber construction: State-of-the-art, gaps and pressing research needs, Journal of Building Engineering. 53 (2022) 104562. https://doi.org/https://doi.org/10.1016/J.JOBE.2022.104562.
59. S.H. Ghaffar, M. Burman, N. Braimah, Pathways to circular construction: An integrated management of construction and demolition waste for resource recovery, J Clean Prod. 244 (2020) 118710. https://doi.org/https://doi.org/10.1016/J.JCLEPRO.2019.118710.
60. M. Spišáková, T. Mandičák, P. Mésároš, M. Špak, Waste Management in a Sustainable Circular Economy as a Part of Design of Construction, Applied Sciences 2022, Vol. 12, Page 4553. 12 (2022) 4553. https://doi.org/10.3390/APP12094553.
61. H. Hasheminasab, S. Hashemkhani Zolfani, M. Kharrazi, D. Streimikiene, Combination of sustainability and circular economy to develop a cleaner building industry, Energy Build. 258 (2022) 111838. https://doi.org/10.1016/J.ENBUILD.2022.111838.
62. T. Adisorn, L. Tholen, T. Götz, Towards a Digital Product Passport Fit for Contributing to a Circular Economy, Energies 2021, Vol. 14, Page 2289. 14 (2021) 2289. https://doi.org/10.3390/EN14082289.

Chapter 7
Circular Supply Chain Management for High-Tech Materials

Abstract In the quick-paced modern world, demand for high-tech materials is rising tremendously. Yet, there is a demand for a more sustainable strategy because the production and consumption of these materials have substantial detrimental effects on the ecosystem. A potential option that makes the shift to a circular economy possible is circular supply chain management. Major adjustments in traceability, collaborations and automation must be made to present supply chain procedures to adopt a close-loop supply chain management system for commodities. By prolonging the useful life of materials via repairs, refurbishment, reprocessing, and reuse, the emphasis is on completing material loops. Manufacturers, suppliers, recyclers, and customers must work together to adopt circular supply chain management in the high-tech materials industry. Moreover, cutting-edge technologies like IoT, blockchain, and big data analytics must be utilized to their fullest potential. This chapter provides an analysis of the merits, challenges, and requirements of the task of the circular supply chain management system for high-tech materials.

Keywords Circularity gap · Agile supply chain · Autonomous systems · High-throughput materials development

7.1 Circular Supply Chain and KPIs

A supply chain is the sequence of processes involved in the production and distribution of a commodity. This includes sourcing raw materials, manufacturing the product, transporting it to the end-user, and ultimately disposing of it [1]. In a linear supply chain, the focus is on producing goods as quickly and cheaply as possible, often resulting in a take-make-waste model where resources are not effectively managed, and waste is generated at every step. In contrast, a circular supply chain aims to minimize waste and maximize resource utilization by keeping products and materials in use for as long as possible [2, 3]. This is achieved through a combination of strategies, such as designing products for reuse and recycling, using

S. Ramakrishna and B. Ramasubramanian, *Handbook of Materials Circular Economy*,
https://doi.org/10.1007/978-981-97-0589-4_7

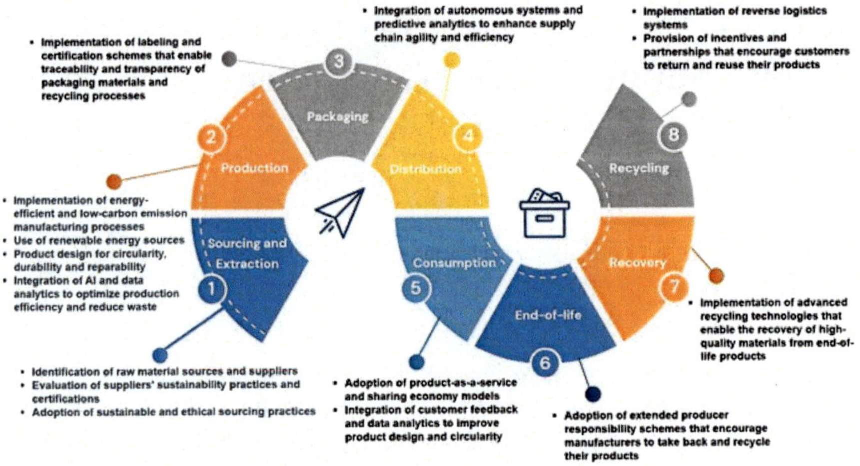

Fig. 7.1 Schematic of updated "Circular Model" from sourcing to recycling

renewable energy sources, and adopting closed-loop systems that keep materials circulating in the economy, as shown in Fig. 7.1.

In the circular economy, all resources are considered valuable, and waste is seen as a potential resource to be used again. The circular supply chain is a critical component of the circular economy and is the means by which goods are brought to the end-user and repurposed through reuse, remaking, and recycling [4, 5]. By creating a closed-loop system, the circular supply chain can minimize waste and ensure that materials and goods are continuously transformed without any negative impact on people or the planet. Ultimately, the circular supply chain is a tool by which everything else in the circular economy links and moves, and it is key to achieving a sustainable and regenerative system [4]. Real-time supply chain models are designed to provide up-to-the-minute information about the status of the supply chain, allowing for quick decision-making and responsive action. These models use real-time data and analytics to monitor inventory levels, demand, and production schedules, allowing businesses to optimize their operations and reduce costs [6–8].

In supply chain management, KPIs (Fig. 7.2) are used to measure the performance of the supply chain and identify areas for improvement. KPIs can be used to measure various aspects of the supply chain, including inventory management, order processing, transportation, and customer service. KPIs are typically quantifiable metrics that are aligned with the goals and objectives of the company [8–11]. They provide valuable insights into the performance of the supply chain, which can be used to identify opportunities for improvement and optimize operations. In the context of the electronic products industry, KPIs are particularly important due to the fast-paced nature of the industry. Consumer demand for electronic products is high, and companies must be able to produce and deliver products to meet this demand quickly and efficiently. KPIs can be used to measure the performance of the supply

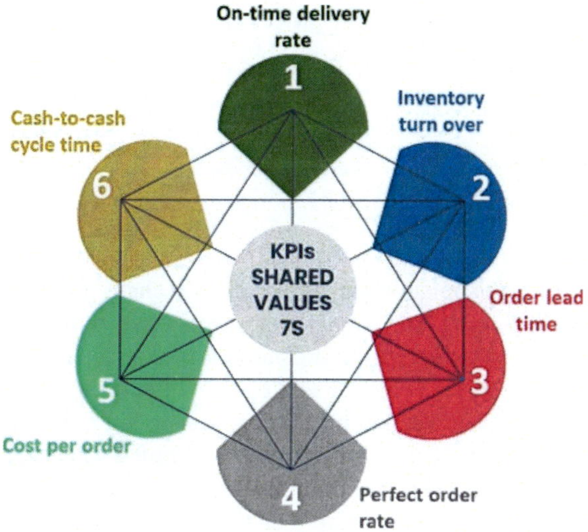

Fig. 7.2 KPIs for sustainable supply chain management

chain in areas such as production efficiency, order processing time, and on-time delivery [10, 11].

Towards achieve successful supply chain management, it is essential to have a deep understanding of the different KPIs that can be used to measure the effectiveness of various processes. The on-time delivery rate KPI, for instance, provides the advantage of ensuring timely delivery of goods or services to maintain customer satisfaction and loyalty [10–13]. Nevertheless, it has the disadvantage of not accounting for external factors that could impact the delivery process. Here, the delivery times are compared to the promised date to track progress. The inventory turnover KPI is another important metric that offers the advantage of preventing inventory shortages and excesses, which could lead to lost sales or increased holding costs [12–15]. However, it does not consider factors such as seasonal demand or product lifespan. Headed for evaluating inventory turnover, the cost of goods sold should be divided by the average inventory value. The order lead time KPI is decisive in identifying areas for improvement in the order processing system, such as delays in order verification or processing [14, 15]. On the other hand, it may not consider other factors that affect order lead time, such as custom orders or backorders. In the direction of ascertaining this KPI, the time it takes from the receipt of an order to the delivery of the order is monitored. The perfect order rate KPI is useful in ensuring that orders are fulfilled correctly, which maintains customer satisfaction and loyalty [1, 15, 16]. Yet, it does not consider factors that may affect the order fulfilment process, such as changes in product specifications or limited product availability. To calculate this KPI, the number of orders with errors or defects is divided by the total number of orders [16]. Finally, the cost per order and cash to cash cycle, KPI identifies areas for cost savings in the order processing system, such as optimizing transportation routes or reducing the number of manual processes. Conversely, it may not consider other factors that impact the cost of order processing, such as changes in product pricing or rising transportation costs [11].

7.1.1 Challenges in Implementing Circular Supply Chain Practices

The implementation of circular supply chain practices presents several challenges that businesses and supply chains must overcome. One of the major challenges is the lack of infrastructure to support these practices [17, 18]. Existing supply chain infrastructure may not be equipped to handle circular practices such as recycling or reusing materials, and investment in new infrastructure may be necessary. In addition to infrastructure challenges, the limited availability of raw materials can pose a significant barrier to the implementation of circular supply chain practices [17]. While circular practices rely on the availability of recycled or repurposed materials, the availability of these materials may be limited in some cases, making it difficult for businesses to adopt circular practices. The complexity of value chains is another challenge that businesses must navigate when implementing circular supply chain practices. Circular practices require coordination across various stakeholders, including suppliers, manufacturers, distributors, and consumers. Coordinating the efforts of these stakeholders can be challenging, particularly in complex and global supply chains [19, 20]. Regulatory and policy barriers can also impact the implementation of circular supply chain practices. For example, regulations related to waste disposal or environmental protection may limit the ability of businesses to implement circular practices [19]. Additionally, policy incentives may be needed to encourage businesses to adopt circular practices, particularly in industries where circular practices are not yet standard. Cultural and behavioural barriers are another challenge that businesses must address when implementing circular supply chain practices. The adoption of circular practices may require changes in behavior and culture, both within organizations and among consumers [1, 21]. This can be a significant challenge, as it requires buy-in and participation from all stakeholders. Furthermore, implementing circular supply chain practices may require significant upfront investment, which can be a financial barrier for businesses, particularly small and medium-sized enterprises [18]. Overcoming this barrier may require innovative financing models, such as public–private partnerships, to provide the necessary resources for businesses to implement circular practices [17]. Finally, data management and technology challenges can also impact the implementation of circular supply chain practices. Circular practices rely on data and technology to track materials and products through the supply chain. However, managing this data and implementing the necessary technology can be a challenge for businesses, particularly those with limited resources [18]. To address this challenge, businesses may need to invest in new technologies and data management systems to enable effective implementation of circular supply chain practices.

7.1.2 *Circularity Gap*

Circularity gap is a measure of the inefficiency that occurs during the recycling process. The principle of recycling is essential for sustainable materials management, but it is important to ensure that the waste produced from recycling is not greater than the recycled product itself [22]. The circularity gap is a critical challenge that exists in many real-world markets today, particularly in industries that rely heavily on resource consumption and waste production. Despite growing awareness of the need for more sustainable materials management practices, the circularity gap remains a significant barrier to achieving a truly circular economy [22]. According to a report by the Ellen MacArthur Foundation, the global circularity gap for materials is estimated to be 91%, meaning that only 9% of materials are recycled and reused, while 91% are lost or wasted. This gap is particularly pronounced in industries such as electronics, where only 17% of materials are recovered and reused, and construction, where only 11% of materials are recycled. This gap is driven by a range of factors, including inefficient recycling processes, lack of infrastructure and investment in recycling and waste management, and poor product design that makes it difficult to recover and reuse materials [22, 23]. In addition, consumer behavior also plays a role, as consumers often fail to properly sort and dispose of materials for recycling.

Data and analytics play a significant role in measuring and monitoring circularity gap in the supply chain. On January 16, 2023, the World Economic Forum released a comprehensive report that offers deep insights into global material flows and crucial recommendations for moving toward a circular economy [24]. In order to help decision-makers make well-informed choices, this report provides them with clear metrics, worldwide data, and measurements of the circular economy [25, 26]. According to Circle Economy's calculations, the world economy is currently only 7.2% circular, a considerable decline from the 9.1% observed in 2018. Just 7.2% of the 100 billion tonnes of virgin materials that are taken from the Planet each year get recycled. The fact that the global economy has used almost as much material in the last 6 years as it did in the entire previous century is disturbing, Fig. 7.3 [26].

7.1.3 *Circularity in Singapore*

Circularity is essential for countries like Singapore, as it can help them reduce their reliance on imports, create new economic opportunities. HP, a leading electronics manufacturer, has implemented a closed-loop recycling system for its ink cartridges. This circular approach has enabled the company to recycle over 3.5 million cartridges annually in Singapore [28]. Similarly, Eco-Wiz, a local food waste management company, has developed a circular model for food waste management at Ang Mo Kio by converting waste into animal feed and fertilizer [29, 30]. The Singaporean government has also been promoting circular practices through various initiatives. In 2020, the National Environment Agency (NEA) launched the Circular Economy

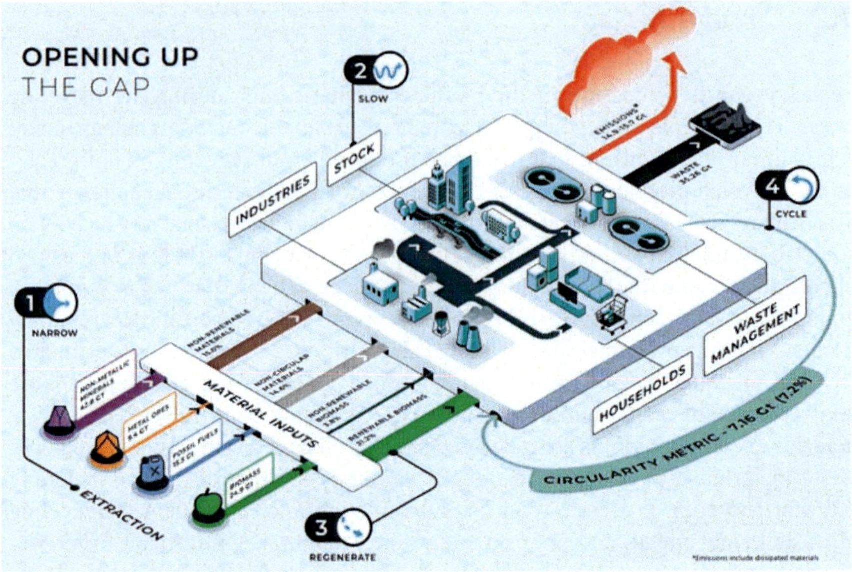

Fig. 7.3 Schematic scenario with inputs and circularity metric [27] https://www.deloitte.com/au/en/issues/climate/global-circularity-gap.html

Industry Transformation Map (ITM) to help businesses and industries in Singapore transition towards a circular economy [29]. The ITM serves as a roadmap for circular economy adoption in sectors such as food, electronics, and packaging. In addition, the government has implemented policies to encourage the adoption of circular practices. The Resource Sustainability Act (RSA) mandates producers and importers of regulated products, such as electronics and packaging, to comply with Extended Producer Responsibility (EPR) requirements. The EPR system makes manufacturers responsible for managing their products' waste at the end of their life cycle, encouraging them to design products that are easier to recycle. Furthermore, the Singaporean government has established infrastructure to support circular practices. The EcoPark, located in Tuas, is a waste management facility that exemplifies circular infrastructure. The facility processes various types of waste, such as plastic, food, and wood, and converts them into new products such as recycled plastic pellets and compost [29].

Singapore is making strides in closing waste loops through circular economy initiatives for plastic, wood, and sludge (Fig. 7.4). Recent developments like a plastics recovery facility and the Mandatory Packaging Reporting (MPR) initiative are promoting sustainability for plastic waste. Local recyclers, upcyclers, and technology providers actively contribute to the plastic waste loop closure. Wood waste is being repurposed by local wood recyclers and upcyclers, while energy recovery solutions like gasification are utilized. Incorporating biochar from recycled wood waste into building materials further supports circularity. Though plans for biogas production

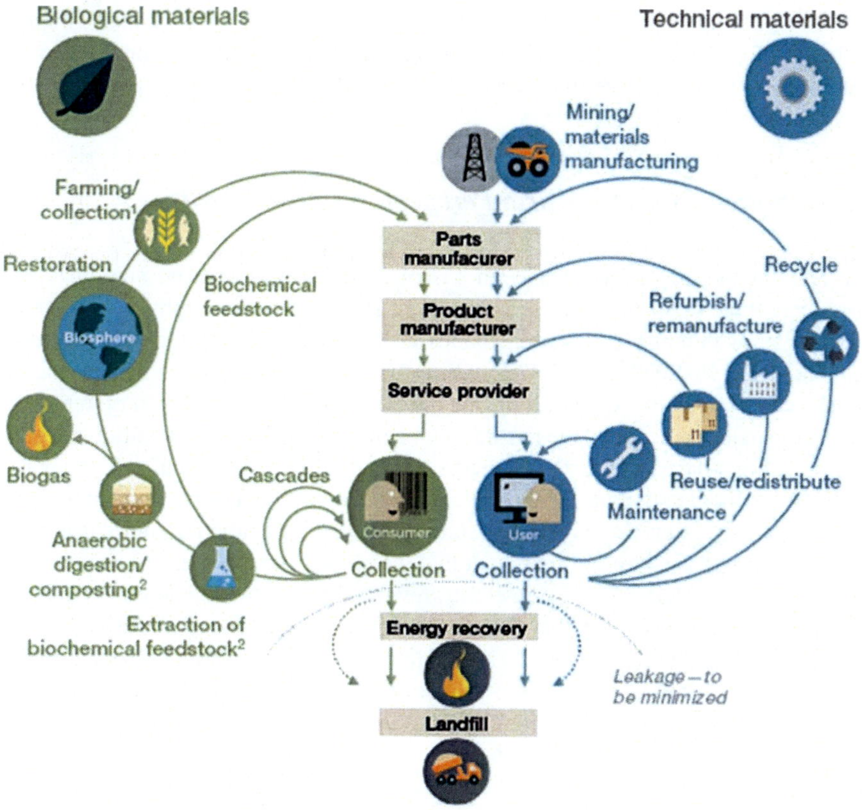

Fig. 7.4 Circularity Model (sourced from the Ellen MacArthur Foundation circular economy team, based on insights from Braungart & McDonough and Cradle to Cradle (C2C) principles) [31]

from sludge show promise, challenges persist in recovering metals and minerals from industrial sludge [28].

The circular economy is implemented in a comprehensive manner by Singapore's zero waste masterplan, which emphasizes the convergence of economic and environmental sustainability along the full value chain. By founding the Sungei Kadut Eco-District, which intends to construct a network of eco-friendly enterprises to take advantage of the opportunities given by the circular economy, Jurong Town Corporation (JTC) has assumed a leadership role in this endeavour. JTC expects that the collaborative products from Sungei Kadut can be utilized as "living labs" to test new concepts because the circular economy demands systemic thinking and creativity. The ecosystem of Singapore will benefit if this legislation is broadly implemented because it will encourage companies to adapt their goods and services to reduce waste and pollution, reuse materials, and regenerate natural systems. By buying only what they need, refraining from overconsumption, and choosing products with circular design, consumers may also contribute to the circular economy. These goods ought to be created from materials that have been recycled, reused, or repurposed and ought to contribute to institutionalized circular processes. Also, consumers want to refrain from wasting resources like food, water, and energy [28, 29, 32].

7.1.4 State-of-the-Art World Perspective in Circularity

The amount of material extracted on a global scale has increased significantly, surpassing three times the amount extracted in 1970 and nearly tripling since 2000 to reach an astounding 100 billion tonnes at this moment. Although the world's population has doubled since 1970, only a 1.7-fold increase in per-person material use can be used to explain all of this development [33]. Despite an increase in per-person material demand, the rise has not been consistent across all nations; high-income nations have seen faster increases in material use than population growth, whilst lower-income nations have experienced the opposite. The metabolism of the global economy is quickening as a result, with material extraction and consumption increasing at an alarming rate, such that the global average does not adequately reflect the entire picture [34, 35]. Over the previous 5 years, the utilization of circular inputs—secondary materials that are reused in the economy—has decreased from 9.1% to 7.2% of all material inputs. This decline isn't just a result of inadequate recycling; it's also a result of rising virgin extraction and stockpiling of resources used to build houses, roads, and durable items. This shows that without a major decrease in material consumption, the global economy is unable to produce a truly closed loop of consumption, and the Circularity Index will keep falling if this trend continues [33].

A quarter of all material inputs are made up of renewable materials, which also include biomass with a high ecological cycling potential (21.2%), which is carbon–neutral, and biomass that is not. Having expanded by a factor of 2.7 over the previous 50 years, biomass currently accounts for around 27% of all material consumption, or 25 billion tonnes annually, and includes everything that is gathered from the earth, from food and feed crops to natural fibers and forest products [36]. While most biomass is classified as renewable, some are not because of the imbalance in the carbon cycle. The cultivation of biomass is a complex process since it frequently involves soil erosion, wetland drainage, and deforestation, all of which have a detrimental impact on biodiversity and carbon sinks and increase emissions. Although carbon neutrality is required for sustainable biomass, it is insufficient on its own. It's also necessary for other nutrients, like phosphorus and nitrogen, to be fully recycled back into the environment or the economy. Ecological cycling has been left out of the calculation of the global Circularity Metric because it is difficult to follow biomass to its final end-of-life stage, which makes it difficult to confirm that the nutrient cycle has closed. However, there are methodological limitations in determining nutrient cycling. Circularity might greatly rise if sustainable biomass handling became commonplace. Metals and non-metallic minerals are among the non-renewable inputs that make up about 15% of all material inputs to the world economy. Throughout the previous 50 years, metal ores have expanded by more than 3.5 times, reaching a total of 9.4 billion tonnes, only one-tenth of the entire extraction. The development of the built environment and manufacturing industries, as well as the switch to clean energy, an essential but resource-intensive process, are all to blame for this substantial increase. In order to achieve clarity, it is necessary to classify the circular process and impacts categories (Fig. 7.5).

The current measurement tool for circularity, the Circularity Metric, solely focuses on the mass-based cycling of materials back into the economy and does not account for their composition, quality, or worth. Consequently, this means that approaches such as "gradual strategies," which prioritize making products last longer, and "slender strategies," which emphasize using fewer resources, are not fully reflected in the metric. Despite the introduction of the full Indicator Framework, which is a positive development in measuring circularity, there are still significant obstacles, such as methodological challenges and data gaps, that prevent a complete picture of the circular economy. Even though including Net additions to stock in the assessment is a first step towards measuring slow strategies, there is still a need for standardized metrics to gain a comprehensive understanding of the circular economy.

The Full Indicator Framework is a comprehensive set of metrics that measures the circularity of an economy, taking into account various factors related to the supply

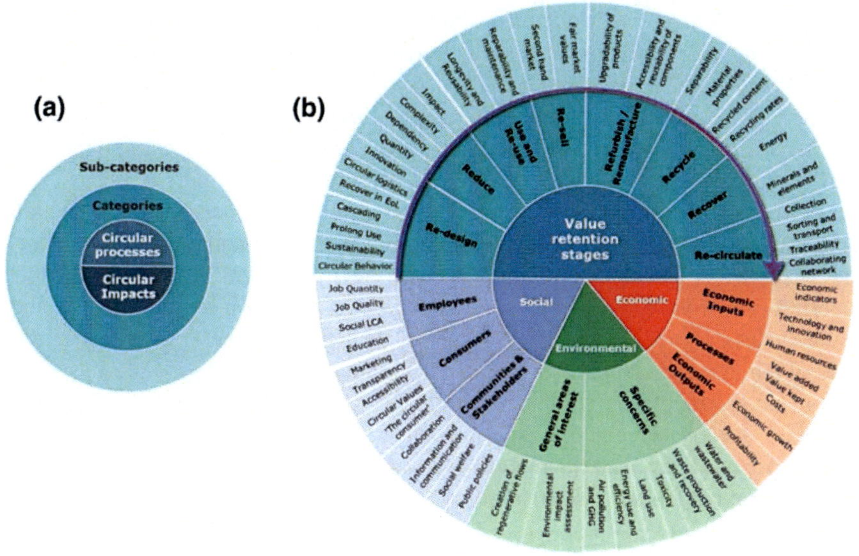

Fig. 7.5 CE Framework Summary: Representation of three layers, b categories, and subcategories, proposed by Garcia-Saravia Ortiz-de-Montellano et al. (a framework composed of two segments. The first segment included circular processes based on product and system value retention on various levels. The second segment measured circular impacts in terms of environmental performance, economic contribution, and social impact. Eight clusters of circular processes were identified, such as redesign, reduce, use and reuse, re-sell, refurbish and remanufacture, recycle, recover, and recirculate) [37]

chain. Developed by the Ellen MacArthur Foundation, the framework is divided into three categories: Enablers, Outputs, and Outcomes. The Enablers category focuses on the fundamental principles of the circular economy, including designing out waste and pollution, keeping products and materials in use, and regenerating natural systems. These guidelines aim to ensure that the supply chain operates in a sustainable and efficient manner. The Outputs category measures the circularity of an economy in practice, assessing the use of renewable and non-toxic materials, waste valorization, product life extension, sharing platforms, product as a service, and resource recovery. These guidelines provide insight into the efficiency and sustainability of production processes, as well as the utilization of products and resources. The Outcomes category measures the benefits of a circular economy for society and the environment. It includes guidelines for job creation and economic development, reduced greenhouse gas emissions, and improved resource security. These guidelines aim to demonstrate the positive impact of circular practices on society and the environment.

7.2 High-Tech Materials

High-tech materials are a class of advanced materials that possess exceptional properties and functionalities, often exceeding those of conventional materials. These materials are highly coveted for their ability to meet complex engineering requirements and provide superior performance across a variety of applications. High-tech materials can be classified based on their composition, structure, and properties. (i) Advanced metals like high-strength alloys, titanium alloys, and superalloys that are frequently employed in aerospace, defence, and medical applications (ii) Advanced composites such as carbon fiber-reinforced polymers (CFRP), glass fiber-reinforced polymers (GFRP), and other fiber-reinforced composites offer high strength-to-weight ratio, stiffness, and durability (iii) Nanomaterials like nanoparticles, nanotubes, and nanowires have unique electronic, optical, and mechanical properties (iv) Advanced ceramics such as oxide ceramics, non-oxide ceramics, and ceramic matrix composites are recognized for their high hardness, wear resistance, and high-temperature stability (v) Advanced polymers like engineering plastics, elastomers, and thermoplastic composites provide high mechanical strength, chemical resistance, and low weight. Lastly, smart materials such as shape memory alloys, piezoelectric materials, and magnetostrictive materials have the ability to respond to external stimuli and alter their properties.

7.2.1 High-Tech Material Sourcing and Production

Materials selection is a critical process that occurs at various stages of product design. The two fundamental steps are screening and ranking (Fig. 7.6), which are typically part of the conceptual design stage. In the initial screening step, a wide range of potential materials is assessed based on their basic properties and characteristics, such as mechanical strength, thermal conductivity, electrical properties, chemical resistance, and availability. This allows for the identification of candidate materials that have the potential to meet the functional requirements and performance criteria of the product. The next step, ranking, involves a more detailed evaluation of the shortlisted materials to assess their suitability for the specific application. Factors considered during ranking include performance under varying conditions, environmental impact, cost, manufacturability, and compliance with relevant regulations and standards. By incorporating screening and ranking into the materials selection process, designers can make informed decisions that lead to the development of successful and innovative products. Alongside materials selection, the product realization process comprises conceptual design, embodiment design, detailed design, and design solution stages to ensure a coherent and effective development process [38].

The type of material and intended usage determine where high-tech materials are sourced. The right selection of materials can significantly enhance the efficiency, reliability, and sustainability of the final product. Figure 7.7 is an example chart for comparing the Youngs modulus and Tensile Strength of materials, engineering materials, including wood, metals and alloys, ceramics and glasses, and polymers and composites, to make informed decisions. Considering the low densities of plant fiber composites and polymer composites, their specific properties become particularly interesting for comparison. In fact, PFRPs demonstrate comparable specific properties to some metals and their alloys due to the latter's higher densities. High-tech material selection within this class of materials is essential to meet specific engineering requirements. Using several synthesis techniques like chemical vapor deposition, sol–gel, microwave-assisted and hydrothermal, nanomaterials can be obtained from commercial manufacturers and laboratory researchers. Researchers can either synthesise nanomaterials themselves, but however for scalability, the commercial manufacturers' customisation services are required. American Elements, Nanoshel, Sigma-Aldrich, and Nanoscale Corporation are only a few companies that manufacture and market nanomaterials. Mxenes, which are two-dimensional materials, are synthesized using chemical etching methods in laboratories, thus limiting their sourcing options. The complex manufacturing processes involved in producing high-tech materials, make scaling up production challenging. Hence, it is necessary to optimize manufacturing processes for efficiency and consistency. Researchers and manufacturers may have to invest in joint specialized equipment and facilities to increase production capacity. For scaling up the production of nanomaterials, techniques such as continuous flow chemical synthesis and electrospinning provide benefits over batch synthesis methods. Yet, they come with high initial investment because of the call for specific tools and knowledge.

Due to their capacity to expand the supplier network, make it easier to access specialized knowledge and expertise, and allow for the customization of materials to meet particular needs, outreach programs, collaborations, and custom synthesis have all demonstrated effectiveness in the sourcing of high-tech materials. At trade exhibitions, conferences, and other events in the industry, outreach programs actively

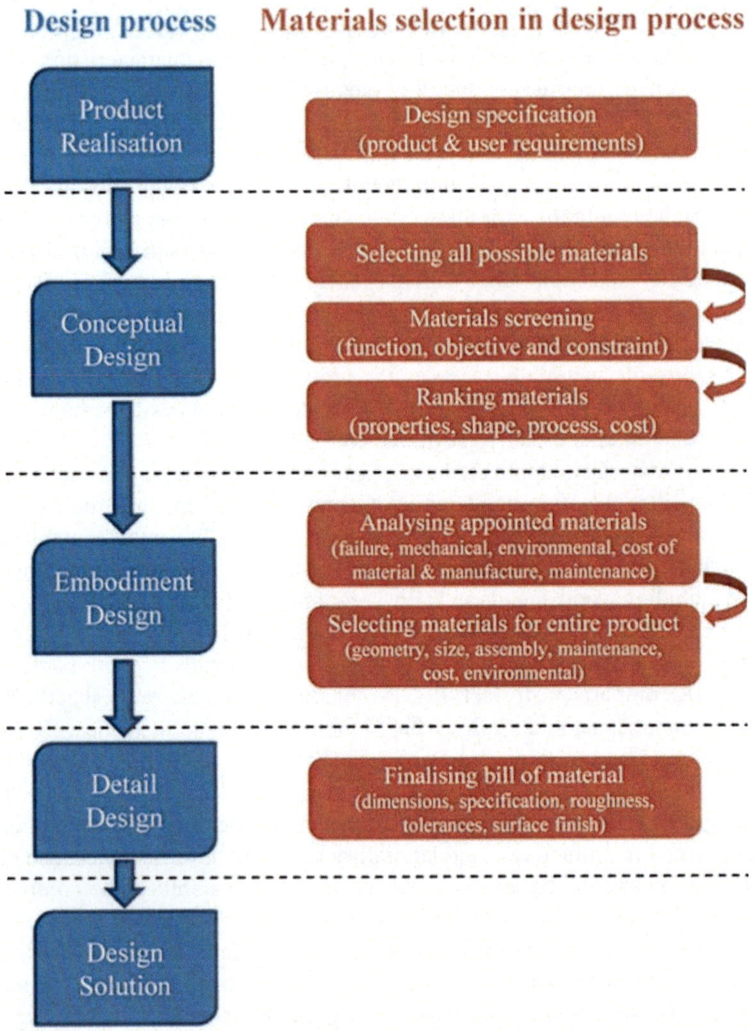

Fig. 7.6 Steps involved for general material selection and design process [38]

seek out potential suppliers and build relationships with them. They may also work with industry groups or governmental organizations to find possible suppliers. On the other hand, collaborations entail working together with other organizations to create new materials or enhance ones that already exist. This could entail working on joint R&D initiatives, establishing joint ventures, or engaging in other kinds of partnerships. Accessing specialist knowledge and skills, cost sharing, and risk reduction are all possible through collaborations. Another strategy for locating high-tech materials is custom synthesis, which entails tight collaboration with vendors to create materials that perfectly satisfy specifications. This can guarantee that the finished

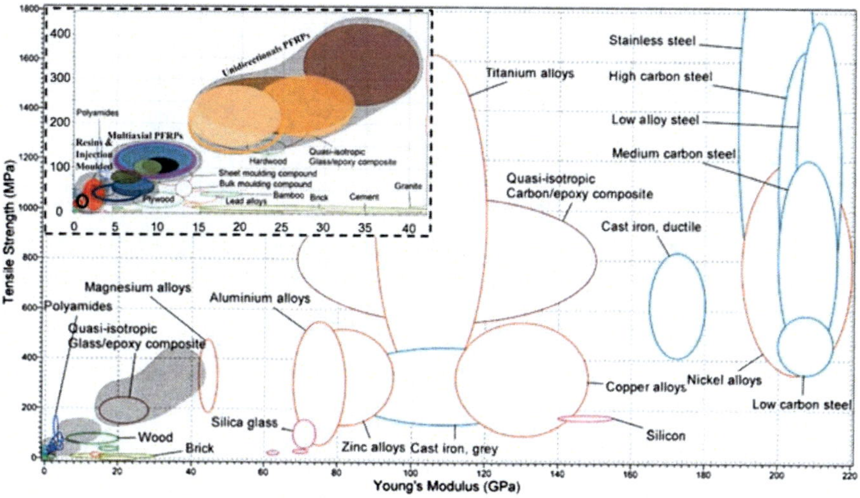

Fig. 7.7 Ashby plots to compare the absolute and specific Young's modulus and tensile strength of plant fiber composites (filled balloons) with other engineering materials (unfilled balloons) [38]

goods satisfy required performance parameters and are suited to the needs of the end customers, and it may entail changing current materials or creating new ones from start. Many standards, tools, and guidelines must be adhered to in order to assure the success of collaborative efforts, shared copyrights, and supply chain standards and policies. Non-disclosure agreements (NDAs) to protect private information, intellectual property agreements to determine ownership and usage rights, ISO certifications for product quality, supply chain management tools like blockchain technology to ensure traceability and transparency, adherence to relevant rules and laws like environmental and safety laws, clear communication channels, and project management structures could all be included.

7.2.2 Supply and Demand of High-Tech Materials

The limited supply of high-tech materials brought on by complicated manufacturing processes and pricey raw materials presents several problems for their manufacture. As a result, it may be challenging to supply the increasing demand for these materials across a variety of sectors, including electronics, energy, aerospace, defence, and biomedicine. The creation of nanomaterials, which are widely sought after for their distinctive features, is one of the major problems. Scaling up the manufacturing procedures necessary to produce nanomaterials, including chemical vapor deposition, can be difficult. Also, expensive raw materials may result in price increases and decreased supply. In a similar vein, smart materials are highly sought after in sectors like aerospace, defence, and biomedicine. Yet, the specific tools and methods required

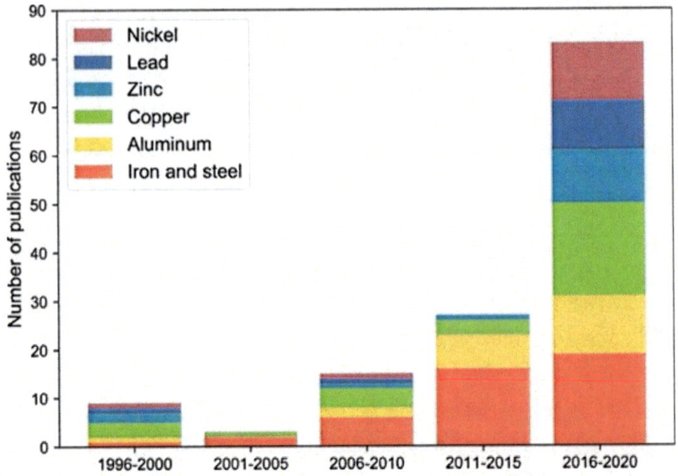

Fig. 7.8 Number of publications on long-term metal outlook, including major metal demand, supply, and environmental implications, along with a comprehensive list of selected articles and covered metals [39]

for their fabrication might make it challenging and costly. For instance, a multi-step process involving shape memory alloy manufacturing necessitates high temperatures and exact control. Researchers and producers are looking into innovative and affordable methods to effectively generate high-tech materials in order to solve these difficulties. They are looking into novel ways to scale up the manufacture of nanomaterials, such as continuous flow chemical synthesis or electrospinning. Furthermore, sophisticated shapes and structures can be made utilizing cutting-edge materials with the development of additive manufacturing techniques like 3D printing.

As an example (Fig. 7.8), over the past 5 years, there has been a significant increase in publications on the long-term outlook for major metal demand, supply, and environmental implications. The largest number of publications focused on iron and steel, followed by copper, aluminum, zinc, nickel, and lead. The focus on these metals suggests that researchers and stakeholders are increasingly recognizing the importance of addressing sustainability issues related to their extraction, processing, use, and disposal. These publications likely explore various aspects, such as resource availability, recycling rates, environmental impacts (e.g., greenhouse gas emissions, water usage, pollution), and potential strategies to ensure a more sustainable metal supply chain.

7.2.3 Global Supply Chain for High-Tech Materials

The acronyms BRICS, CIVETS, and MINT are used to refer to emerging economies that are expected to have significant impacts on the global economy. The BRICS stands for Brazil, Russia, India, China, and South Africa, and are anticipated to become dominant players in the global economy due to their large populations, abundant natural resources, and high economic growth rates. CIVETS stands for Colombia, Indonesia, Vietnam, Egypt, Turkey, and South Africa, and are also emerging economies with strategic locations, growing middle classes, and high levels of economic growth. MINT stands for Mexico, Indonesia, Nigeria, and Turkey, and these countries are characterized by their large populations, growing middle classes, and potential for economic growth. These groups are often studied and compared based on their economic growth, trade relations, and potential for investment and collaboration.

Nanomaterials and advanced smart materials are two key categories of advanced materials that are becoming more and more integrated into a variety of sectors, including electronics, energy, and the healthcare industry. The nanotechnology in energy applications market is projected to grow at a CAGR of 15.0% from 2023 to 2028, reaching $18.8 billion by 2028, compared to $9.3 billion in 2023 [40]. China has been the top BRICS nation provider to the world's supply chain for sophisticated materials, particularly in the field of nanomaterials. China, with a market share of over 35%, is the world's largest producer and user of nanomaterials, according to 2021 research by the Royal Society of Chemistry. Colombia has built a number of centers and research groups for nanotechnology, as well as made investments in the field [41, 42]. Vietnam has also been boosting its investment in nanomaterials research and development, and the country is expected to have the Asia–Pacific region's fastest-growing market for smart materials. Mexico and Turkey have been recognized as MINT nations that contribute significantly to the world's supply chain for sophisticated materials. With a significant focus on applications in the fields of energy and the environment, Mexico has been investing in the research and development of nanoscale materials. Turkey has developed a number of centers and research groups in the field of nanotechnology and has been increasing its investment in this area [40, 43].

High-tech materials are mostly produced and supplied by nations like the United States, Japan, Germany, and South Korea. Yet, because to their impressive investments in R&D and expanding production capacities, rising economies like China, Taiwan, and Singapore are quickly emerging as major participants in this sector. The development of advanced materials that are more effective, economical, and environmentally friendly is the main goal of transformational growth for high-tech materials, which is centered on innovation and new technologies. It is also essential to create new products and applications that make use of these materials for improved performance and usefulness. Country strategies for high-tech materials comprise the establishment of policies and initiatives, such as investments in R&D, innovation hubs, and promotion of entrepreneurship and startups in the industry, to promote the industry's growth. Also, nations concentrate on enhancing their manufacturing capacities and forming alliances with other nations to promote trade and cooperation. Since it speeds up research and development through the sharing of knowledge, skills, and resources, international collaboration is an essential component of the high-tech materials sector. International cooperation can also be used to set global norms and rules that will support high-tech material production and use that is both secure and green. The high-tech materials sector must develop strategies and procedures to reduce risks and adjust to shifting market conditions. To reduce geopolitical risks, this entails diversifying supply chains, making investments in research and development to provide substitute materials, and forming alliances with other nations.

7.3 Key Trends for High-Tech Materials

The business ecosystem in which the high-tech materials sector operates is intricate and includes a wide range of players, including suppliers, manufacturers, customers, regulators, and rivals. Interdependence, cooperation, and innovation are the hallmarks of this ecosystem, in which each participant is essential to the success of the sector. Businesses must prioritize innovation, efficient supply chain management, and regulatory compliance if they want to stay competitive in this environment. Staying ahead of the competition through innovation is one of the major problems that high-tech materials manufacturers face [44, 45]. This entails creating novel, technologically advanced materials that can adapt to the changing needs of the market while also being more effective and affordable. Businesses also need to properly manage their supply chains to guarantee a steady supply of components, completed goods, and commodities. Opportunities for firms to access new markets and resources have been made possible by globalization [44–46].

7.3.1 Prospects for Globalization for High-Tech Materials

The high-tech materials sector is substantially impacted by demographic changes and population expansion. There is a rising need for high-tech materials in industries including infrastructure, transportation, and energy as the world's population continues to expand. Meanwhile, the need for novel materials and goods that might enhance quality of life is driven by shifting demographics like an aging population. Another crucial issue for the sector is gender diversity [47, 48]. Although being historically controlled by men, efforts to encourage gender diversity can open up fresh viewpoints, boost productivity, and advance equity. The industry has created new materials and technology to meet consumers' evolving preferences as their demands for environmentally friendly and socially conscious products have increased. Demand for high-tech materials has also increased due to urbanization, which has sparked the creation of affordable and environmentally beneficial materials. Sustainability, innovation, and connectivity are the defining characteristics of mobility in the twenty-first century, and they have sparked the creation of new materials and technologies that increase fuel economy, lower emissions, and increase safety. Contrarily, the usage of cutting-edge technology like artificial intelligence, machine learning, and the Internet of Things is altering the business thanks to digital culture. The industry has to develop new materials that enable ubiquitous intelligence, or the integration of cutting-edge technologies into commonplace things and environments, in order to handle this challenge [47].

The industry is also utilizing the field of bionics, which involves the study of biological systems and the creation of materials with qualities similar to those of such systems. The industry is responding by creating new materials and technologies that lessen dependency on non-renewable resources and boost energy efficiency. Resource shortages and energy transitions provide substantial problems. Last but not least, one of the biggest problems the sector is dealing with is climate change [49]. The sector must create new products and processes that lower greenhouse gas emissions, boost energy effectiveness, and support sustainability. To do this, the sector must control the negative environmental effects of material creation and disposal while lowering the sector's overall carbon footprint [48].

7.3.2 Trajectory of Supply Chains for the Future

Supply chains are becoming more automated, intelligent, and efficient with the introduction of new technologies like artificial intelligence, machine learning, and blockchain. These innovations are making it possible to build digital supply chains that offer end-to-end visibility, transparency, and traceability in real time. Moreover, the Internet of Things (IoT) integration is promoting the creation of smart supply chains that can optimize logistics, monitor and manage inventory, and enable predictive maintenance. Supply chain managers can acquire a comprehensive understanding of their operations by utilizing IoT devices, such as sensors and RFID tags, which can assist them in making data-driven choices and enhancing performance [50, 51]. Moreover, supply chain processes are growing progressively automated and robotic, which increases efficiency, lowers costs, and increases safety. Sorting, packing, and transportation-related operations are handled by automation, while picking, loading, and palletizing are handled by robotics. While enhancing the efficiency and accuracy of supply chain operations, these technologies can help decrease the likelihood of workplace accidents and injuries. Additionally, improvements in supply chain analytics are enabling businesses to better understand their operations, spot bottlenecks, and streamline their workflows. Supply chain managers can use data analytics to acquire real-time insights into the performance of their supply chains, which can help them spot areas for improvement and provide better results [52, 53].

7.4 Emerging Technologies for Supply Chain Management

End-to-end design, high visibility, transparency, alignment, and tight collaboration are characteristics of agile supply chains. These characteristics are clear in the planning, design, and implementation processes. The performance and cost of agile supply networks are significantly impacted by their great adaptability. Businesses with agile supply chains can efficiently allocate inventory across distribution networks and incur fewer costs for transportation, handling, production, or service provision. The development of agile supply chains depends on nine key priority areas that have been identified by the German Association of Logistics (BVL) [54, 55]. Transparency is one of these areas of emphasis, and it requires tracking changes in consumer expectations, market dynamics, and technological advancements. Companies can modify their tactics as necessary thanks to BVL analysis. Conscious buy-or-make decisions are a different area of concentration. Retaining essential skills and services is advised for businesses. At least two suppliers or service providers should be accessible if this is not practicable. Determining the effectiveness and durability of the supply chain also heavily depends on supplier selection. Collaboration is a key success factor in joint development, which is essential for setting up the supply chain. Better alignment of all participants within the supply chain ecosystem is achieved by including suppliers and service providers in design considerations of new goods and processes. To organize supplier and service provider participation in the creation of products and services, independent development units can also be established. Talent pools and competency networks must be established in order to construct an entire human resources (HR) operation. Based on their abilities and desire to pitch in, HR pools give flexibility to assign and engage staff where needed. Another crucial area that encourages accountability and shared responsibility across teams is dynamic quality assurance. Also, since they fix a goal on their own, dynamic, self-governing project groups are required to construct a roadmap. Developing and maintaining agile supply chains requires expertise, which is shared and transferred through coaching, theme alliances and cooperation, and an open feedback culture at the multinational logistics business Cargo. Employees may engage rapidly both within and outside of their areas of duty and authority thanks to an integrated platform with a compendium that is available to all logged-in users. They can write, update, or do research on articles, papers, reports, IT product user manuals, and newspaper articles. The site also includes lists for real estate, office supplies, and corporate bids, as well as a dynamic subscription registry and lifecycle idea management [50, 54, 55].

Advanced data management techniques are crucial to the success of agile supply chains. Internet Data Center (IDC) data researchers estimate that between 2005 and 2020, the digital universe will have grown by a factor of 300, from 130 exabytes to 40,000 exabytes. Its spread has been aided by the data produced by numerous sources, including cell phones, email, and other digital devices [56]. Around 300 million new bits of information are created every day by parcel delivery activities, which puts large data volumes under management for logistics service providers like DHL. In this situation, contemporary logistics, or e-logistical, which is dependent on data management, has evolved into a crucial part of logistics operations. Data gathering, analysis, and storage now involve sharing in addition to internal usage. Transport and logistics firms are now able to reduce risks, identify fraud, plan workforce and capacity, and improve performance thanks to the availability of enormous volumes of data [54, 56]. The On-Road Integration Optimization and Navigation (ORION) initiative by UPS serves as an illustration of how data analysis and smart technology may result in considerable fuel and cost savings. Big data utilization may significantly affect the triple bottom line, which includes economic, social, and environmental aspects. The success of agile supply chains depends on improved data management techniques, and logistics service providers must adjust to the evolving technology environment to stay competitive in the market.

The most successful and economical supply chain is an autonomous system that is intelligent, secure, safe, and flexible. In order to make informed judgments, learn from the past, and navigate through a highly complex environment, an autonomous supply chain must take into consideration all available data. Since the early 1950s, advanced navigation transportation and vehicles have been employed in production and logistics. Nowadays, they are broadly applied in the supply chain market, where they assist to improve material flow and save costs. Machinery for driverless loading and transportation aids in lowering costs and distribution network mistakes. Convoy battalions that drive closely together and communicate wirelessly have been successfully tested on public roadways, and driverless transport is going to transform container ships. The last mile, or last delivery phase of the distribution chain, is the most complicated, particularly in urban settings. Businesses are developing the algorithms and intellect machine learning necessary for the autonomous world. With raw materials being mined in automated mines, transferred to smart factories, and finally supplied to clients through underground, vehicle battalions, long-distance drones, or driverless distribution facilities, the distribution network of the tomorrow is almost complete. Drones, robots, urban canals, or even individual self-driving automobiles will deliver orders to customers. Logistics and infrastructure are becoming more important as society evolves toward an unmanned era.

The delivery of the things we use on a daily basis is made possible by the global supply chain, which is essential in maintaining the contemporary economy and society. Expertise in procuring, purchasing, shipping, and distribution, as well as excellent international stewardship to maintain a high quality of life for consumers and residents, are needed to properly develop and operate this ecosystem [57]. To guide the ecosystem in the correct path, Chief Supply Chain Officers (CSCOs) must demonstrate strong leadership. CSCOs must continuously upgrade their expertise to create new supply chain administration models as technology evolves and society becomes more linked. The supply chain sector now encompasses a refreshed set of fundamental circularity ideas about the extraction, transportation, manufacture, and reuse rather than being restricted to technical and operational procedures. All items and supplies will be traceable in the future, while those that lack an open "tracking code" will be subject to further inspection, which will obstruct their flow through the ecosystem. As explicit as the point of origin, information on product components, life cycles, and reprocessing will be provided. This level of openness will continuously disseminate information on the components, finished goods, manufacturing procedures, as well as usage, upkeep, maintenance, disposal, and recycling. In the context of its Circular Economy Action Plan, the European Union also pioneered in the development of product passports [57, 58]. A digital product passport has been suggested by the EU that would contain details on a product's environmental effect, place of manufacture, materials used, and other pertinent information throughout its full life cycle. This program aims to increase product sustainability and give customers greater choice over their purchase decisions [59–61].

7.5 Circularity Approaches in Supply Chain

As per Harvard Business school, in managing sustainability within their building material supply networks, multinational corporations (MNCs) adopt four distinct approaches [62, 63]. The first is the direct approach, where MNCs establish specific social and environmental targets for their primary suppliers, promoting sustainable practices and diversity. Regular evaluations and opportunities for lower-tier suppliers to network are utilized to ensure compliance and foster inclusivity. The second approach, known as the indirect approach, involves MNCs delegating the responsibility of lower-tier supplier sustainability management to their first-tier suppliers. Through training programs and incentives, these first-tier suppliers are encouraged to adopt and implement sustainable practices. Preferred-supplier initiatives further encourage peer learning and the adoption of sustainability standards [62]. The third approach, the collective approach, sees MNCs collaborating with competitors and major suppliers to develop comprehensive industrywide sustainability standards, assessment tools, and training programs. Industry associations, like the Responsible Business Alliance (RBA), play a crucial role in facilitating standardized audits and assessments. Lastly, MNCs adopt a global approach, which involves partnering with international organizations and non-governmental organizations (NGOs), including the United Nations Global Compact and the Carbon Disclosure Project (CDP). This collaboration aims to advance shared sustainability objectives. The CDP's Supply Chain Program is used as a platform to encourage suppliers to disclose information about their environmental practices and make improvements accordingly [62, 63].

Effective collaboration with suppliers is essential for integrating supply chains and effectively managing risks and impacts, enabling organizations to seize value creation opportunities. Such collaboration requires aligning strategic goals, engaging cross-functional teams, establishing governance, fostering communication, and sharing value among stakeholders. By aligning suppliers with critical objectives, organizations can drive innovation, enhance quality, reduce costs, and improve customer satisfaction [64, 65]. Circular economy initiatives play a vital role in waste reduction and sustainability by promoting recycling and reducing reliance on new resources. Implementing eco-friendly sourcing strategies throughout the supply chain, including ethical policies, supports long-term sustainability and risk reduction. End-to-end supply chain visibility allows organizations to closely monitor components, materials, and finished goods, empowering them to make informed decisions and respond promptly to changing circumstances. Sustainability compliance strategies ensure that ethical and environmental standards are upheld by all stakeholders. Embracing digital technologies enhances operational efficiency, boosts supply chain transparency, and ensures accuracy in operations. Implementing transportation optimization strategies minimizes costs and emissions, enabling organizations to maintain a competitive edge while promoting sustainability.

Despite the progress made through these approaches, there remains ample room for improvement in enhancing sustainability practices beyond the first-tier suppliers in the building material supply chains. Continued efforts and initiatives are necessary to drive meaningful change and create a more sustainable and environmentally responsible supply network for the future.

7.6 Conclusion

In summary, Circular Supply Chain Management (CSCM) adoption for high-tech materials has the ability to close the circularity gap and offer a long-term solution for the effective use and recovery of resources. CSCM requires an agile supply chain that can adapt to the shifting demands of the market and the product's accessibility. The procedure may be streamlined and the possibility of human error decreased with the use of autonomous systems. High-throughput materials development can indeed simplify the search for and develop novel materials that are more ecologically friendly and resilient. For the success of CSCM, the usage of product passports can assist assure transparency and traceability across the supply chain. The old linear model of production and consumption should undergo a paradigm change with the introduction of CSCM for high-tech materials. Yet for CSCM to be successfully implemented, all relevant parties—producers, consumers, regulators, and policymakers—must work together. Together, we can reduce the circularity gap and build a responsible future that is highly self-sustaining.

Activity: MCQs

1. What is the primary objective of circular supply chain management for high-tech materials?

   (a) To increase waste generation
   (b) To promote linear material flow
   (c) To reduce resource consumption and waste
   (d) To encourage single-use materials

2. Which of the following is NOT a characteristic of circular supply chain management for high-tech materials?

   (a) Extending the product lifespan through maintenance and repair

(b) Prioritizing one-time use of materials
(c) Reusing and recycling materials in the supply chain
(d) Minimizing resource extraction through material recovery

3. How does circular supply chain management contribute to sustainability?

(a) By increasing resource consumption
(b) By promoting planned obsolescence
(c) By reducing waste generation and promoting material reuse
(d) By encouraging landfilling of high-tech materials

4. Which circular practice can be applied to high-tech materials to achieve circular supply chain management?

(a) Rapid disposal of electronic waste
(b) Reusing and refurbishing high-tech components
(c) Promoting single-use materials in the supply chain
(d) Exporting high-tech materials to developing countries

5. Which of the following is an example of circular supply chain management for high-tech materials?

(a) Dumping electronic waste in oceans
(b) Incinerating old electronic components
(c) Recovering valuable materials from discarded electronics
(d) Using non-recyclable materials in high-tech products

6. What is a key advantage of adopting circular supply chain management for high-tech materials?

(a) Increased resource extraction
(b) Higher waste generation
(c) Greater cost savings and sustainability
(d) Limited material recovery opportunities

7. Which circular approach in high-tech supply chains involves using recycled materials in manufacturing processes?

(a) Landfilling of electronic waste
(b) Incineration of old high-tech components
(c) Reusing and remanufacturing electronic parts
(d) Rapid disposal of high-tech materials

8. How does circular supply chain management contribute to waste reduction?

(a) By promoting planned obsolescence of high-tech products
(b) By encouraging single-use materials in the supply chain
(c) By promoting material reuse and recycling within the supply chain
(d) By ignoring the environmental impact of high-tech materials

9. Which of the following is a circular practice in high-tech supply chains that prioritizes sustainability?
   (a) Using non-recyclable materials in high-tech products
   (b) Rapidly disposing of electronic waste
   (c) Adopting eco-friendly and recyclable materials
   (d) Ignoring the environmental impact of high-tech production
10. Circular supply chain management for high-tech materials aims to achieve:
   (a) Increased waste generation
   (b) Greater resource consumption
   (c) Sustainable and efficient resource utilization
   (d) Escalating material extraction

Answers:

1. (c) To reduce resource consumption and waste
2. (b) Prioritizing one-time use of materials
3. (c) By reducing waste generation and promoting material reuse
4. (b) Reusing and refurbishing high-tech components
5. (c) Recovering valuable materials from discarded electronics
6. (c) Greater cost savings and sustainability
7. (c) Reusing and remanufacturing electronic parts
8. (c) By promoting material reuse and recycling within the supply chain
9. (c) Adopting eco-friendly and recyclable materials
10. (c) Sustainable and efficient resource utilization

References

1. Medina G, Thomé K (2021) Transparency in Global Agribusiness: Transforming Brazil's Soybean Supply Chain Based on Companies' Accountability. Logistics 2021, Vol 5, Page 58 5:58. https://doi.org/10.3390/LOGISTICS5030058
2. Khadke S, Gupta P, Rachakunta S, et al (2021) Efficient Plastic Recycling and Remolding Circular Economy Using the Technology of Trust–Blockchain. Sustainability 2021, Vol 13, Page 9142 13:9142. https://doi.org/10.3390/SU13169142
3. Zhang T, Liu H, Han X, et al (2023) Improvement of carbon dioxide electroreduction by crystal surface modification of ZIF-8. Dalton Transactions. https://doi.org/https://doi.org/10.1039/D3DT00185G
4. Koberg E, Longoni A (2019) A systematic review of sustainable supply chain management in global supply chains. J Clean Prod 207:1084–1098. https://doi.org/https://doi.org/10.1016/J.JCLEPRO.2018.10.033
5. Ghadge A, Wurtmann H, Seuring S (2019) Managing climate change risks in global supply chains: a review and research agenda. https://doi.org/10.1080/00207543.2019.1629670 58:44–64. https://doi.org/10.1080/00207543.2019.1629670
6. Onat NC, Kucukvar M (2020) Carbon footprint of construction industry: A global review and supply chain analysis. Renewable and Sustainable Energy Reviews 124:109783. https://doi.org/https://doi.org/10.1016/J.RSER.2020.109783

7. Guan D, Wang D, Hallegatte S, et al (2020) Global supply-chain effects of COVID-19 control measures. Nature Human Behaviour 2020 4:6 4:577–587. https://doi.org/10.1038/s41562-020-0896-8
8. Ali I, Alharbi OML (2020) COVID-19: Disease, management, treatment, and social impact. Science of the Total Environment 728:. https://doi.org/10.1016/J.SCITOTENV.2020.138861
9. Uwizeye A, de Boer IJM, Opio CI, et al (2020) Nitrogen emissions along global livestock supply chains. Nature Food 2020 1:7 1:437–446. https://doi.org/10.1038/s43016-020-0113-y
10. Flynn B, Cantor D, Pagell M, et al (2021) From the Editors: Introduction to Managing Supply Chains Beyond Covid-19 - Preparing for the Next Global Mega-Disruption. Journal of Supply Chain Management 57:3–6. https://doi.org/https://doi.org/10.1111/JSCM.12254
11. Chang Y, Iakovou E, Shi W (2019) Blockchain in global supply chains and cross border trade: a critical synthesis of the state-of-the-art, challenges and opportunities. https://doi.org/10.1080/00207543.2019.1651946 58:2082–2099. https://doi.org/10.1080/00207543.2019.1651946
12. Litke A, Anagnostopoulos D, Varvarigou T (2019) Blockchains for Supply Chain Management: Architectural Elements and Challenges Towards a Global Scale Deployment. Logistics 2019, Vol 3, Page 5 3:5. https://doi.org/10.3390/LOGISTICS3010005
13. Bhaskar S, Tan J, Bogers MLAM, et al (2020) At the Epicenter of COVID-19–the Tragic Failure of the Global Supply Chain for Medical Supplies. Front Public Health 8:821. https://doi.org/https://doi.org/10.3389/FPUBH.2020.562882/BIBTEX
14. Kuruvilla S, Li C (2021) Freedom of Association and Collective Bargaining in Global Supply Chains: A Research Agenda. Journal of Supply Chain Management 57:43–57. https://doi.org/https://doi.org/10.1111/JSCM.12259
15. Anner M (2019) Predatory purchasing practices in global apparel supply chains and the employment relations squeeze in the Indian garment export industry. Int Labour Rev 158:705–727. https://doi.org/https://doi.org/10.1111/ILR.12149
16. Gardner TA, Benzie M, Börner J, et al (2019) Transparency and sustainability in global commodity supply chains. World Dev 121:163–177. https://doi.org/https://doi.org/10.1016/J.WORLDDEV.2018.05.025
17. Tsanakas JA, Heide A van der, Radavičius T, et al (2020) Towards a circular supply chain for PV modules: Review of today's challenges in PV recycling, refurbishment and re-certification. Progress in Photovoltaics: Research and Applications 28:454–464. https://doi.org/https://doi.org/10.1002/PIP.3193
18. Frei R, Jack L, Krzyzaniak SA (2020) Sustainable reverse supply chains and circular economy in multichannel retail returns. Bus Strategy Environ 29:1925–1940. https://doi.org/https://doi.org/10.1002/BSE.2479
19. Hossain MU, Ng ST, Antwi-Afari P, Amor B (2020) Circular economy and the construction industry: Existing trends, challenges and prospective framework for sustainable construction. Renewable and Sustainable Energy Reviews 130:109948. https://doi.org/https://doi.org/10.1016/J.RSER.2020.109948
20. Jia F, Yin S, Chen L, Chen X (2020) The circular economy in the textile and apparel industry: A systematic literature review. J Clean Prod 259:120728. https://doi.org/https://doi.org/10.1016/J.JCLEPRO.2020.120728
21. Yadav G, Luthra S, Jakhar SK, et al (2020) A framework to overcome sustainable supply chain challenges through solution measures of industry 4.0 and circular economy: An automotive case. J Clean Prod 254:120112. https://doi.org/10.1016/J.JCLEPRO.2020.120112
22. Guerra BC, Shahi S, Molleai A, et al (2021) Circular economy applications in the construction industry: A global scan of trends and opportunities. J Clean Prod 324:129125. https://doi.org/https://doi.org/10.1016/J.JCLEPRO.2021.129125
23. James AM, Reitsma L, Aftab M (2019) The Design Journal An International Journal for All Aspects of Design ISSN: (Print) (Online) Journal homepage: https://www.tandfonline.com/loi/rfdj20 Bridging the double-gap in circularity. Addressing the intention-behaviour disparity in fashion Bridging the double-gap in circularity. Addressing the intention-behaviour disparity in fashion. 10–12. https://doi.org/10.1080/14606925.2019.1595407
24. B E H IN D TH E COV E R

25. Antwi-Afari P, Ng ST, Hossain MU (2021) A review of the circularity gap in the construction industry through scientometric analysis. J Clean Prod 298:126870. https://doi.org/https://doi.org/10.1016/J.JCLEPRO.2021.126870
26. Aguilar-Hernandez GA, Sigüenza-Sanchez CP, Donati F, et al (2019) The circularity gap of nations: A multiregional analysis of waste generation, recovery, and stock depletion in 2011. Resour Conserv Recycl 151:104452. https://doi.org/https://doi.org/10.1016/J.RESCONREC.2019.104452
27. The Global Circularity Gap. https://www.deloitte.com/au/en/issues/climate/global-circularity-gap.html. Accessed 14 Dec 2023
28. Carrière S, Weigend Rodríguez R, Pey P, et al (2020) Circular cities: the case of Singapore. Built Environment Project and Asset Management 10:491–507. https://doi.org/https://doi.org/10.1108/BEPAM-12-2019-0137/FULL/XML
29. Arora M, Raspall F, Cheah L, Silva A (2019) Residential building material stocks and component-level circularity: The case of Singapore. J Clean Prod 216:239–248. https://doi.org/https://doi.org/10.1016/J.JCLEPRO.2019.01.199
30. Eco-Wiz – Zero Waste SG. http://www.zerowastesg.com/tag/eco-wiz/. Accessed 20 Jul 2023
31. The Butterfly Diagram: Visualising the Circular Economy. https://www.ellenmacarthurfoundation.org/circular-economy-diagram. Accessed 14 Dec 2023
32. Cowen T (2023) Singapore facts of the day, shareholder state edition
33. Fathi BM, Ansari A, Ansari A (2000) Threats of Internet-of-Thing on Environmental Sustainability by E-Waste. https://doi.org/10.3390/su141610161
34. Modarress Fathi B, Ansari A, Ansari A (2022) Threats of Internet-of-Thing on Environmental Sustainability by E-Waste. Sustainability 2022, Vol 14, Page 10161 14:10161. https://doi.org/10.3390/SU141610161
35. Ashrafuzzaman M, Santos FD, Dias JM, Cerdà A (2022) Dynamics and Causes of Sea Level Rise in the Coastal Region of Southwest Bangladesh at Global, Regional, and Local Levels. J Mar Sci Eng 10:779. https://doi.org/https://doi.org/10.3390/JMSE10060779/S1
36. Thunman H, Gustavsson C, Larsson A, et al (2019) Economic assessment of advanced biofuel production via gasification using cost data from the GoBiGas plant. Energy Sci Eng 7:217–229. https://doi.org/https://doi.org/10.1002/ESE3.271
37. Garcia-Saravia Ortiz-de-Montellano C, van der Meer Y, van der Meer YvonnevanderMeer Y (2022) A Theoretical Framework for Circular Processes and Circular Impacts Through a Comprehensive Review of Indicators. Global Journal of Flexible Systems Management 2022 23:2 23:291–314. https://doi.org/10.1007/S40171-022-00300-5
38. Shah DU (2020) Materials Selection Charts for Designing Products With Biocomposites. Encyclopedia of Renewable and Sustainable Materials: Volume 1-5 1–5:768–780. https://doi.org/https://doi.org/10.1016/B978-0-12-803581-8.11658-3
39. Watari T, Nansai K, Nakajima K (2021) Major metals demand, supply, and environmental impacts to 2100: A critical review. Resour Conserv Recycl 164:105107. https://doi.org/https://doi.org/10.1016/J.RESCONREC.2020.105107
40. Nanotechnology Markets Size Share Forecast & Analysis Reports. https://www.bccresearch.com/market-research/nanotechnology. Accessed 22 Jul 2023
41. Nguyen PH, Nguyen TL, Le HQ, et al (2023) How Does the Competitiveness Index Promote Foreign Direct Investment at the Provincial Level in Vietnam? An Integrated Grey Delphi–DEA Model Approach. Mathematics 2023, Vol 11, Page 1500 11:1500. https://doi.org/10.3390/MATH11061500
42. Nanotechnology Market Top Companies by Revenue | Nanotech industry key companies by 2028. https://www.emergenresearch.com/blog/nanotechnology-companies-to-watch-out-for-in-the-near-future. Accessed 22 Jul 2023
43. Nanomedicine Market Size, Share, Revenue, Trend, Forecast 2030. https://www.reportsanddata.com/report-detail/nanomedicine-market. Accessed 22 Jul 2023
44. Liu K, Wang X, Yan Y (2022) Network Analysis of Industrial Symbiosis in Chemical Industrial Parks: A Case Study of Nanjing Jiangbei New Materials High-Tech Park. Sustainability 2022, Vol 14, Page 1381 14:1381. https://doi.org/https://doi.org/10.3390/SU14031381

45. Stel'makh SA, Shcherban' EM, Beskopylny A, et al (2022) Development of High-Tech Self-Compacting Concrete Mixtures Based on Nano-Modifiers of Various Types. Materials 2022, Vol 15, Page 2739 15:2739. https://doi.org/10.3390/MA15082739
46. Murtaja Y, Lapčík L, Lapčíková B, et al (2022) Intelligent high-tech coating of natural biopolymer layers. Adv Colloid Interface Sci 304:102681. https://doi.org/https://doi.org/10.1016/J.CIS.2022.102681
47. Holzmann R, Slačík T, Wörz J (2023) Conference on European Economic Integration (CEEI) 2021: Recalibrating tomorrow's global value chains – prospects for CESEE. Empirica 50:1–5. https://doi.org/10.1007/S10663-022-09565-6/FIGURES/2
48. Du L, Jiang H, Adebayo TS, et al (2022) Asymmetric effects of high-tech industry and renewable energy on consumption-based carbon emissions in MINT countries. Renew Energy 196:1269–1280. https://doi.org/https://doi.org/10.1016/J.RENENE.2022.07.028
49. Popkova EG (2022) INTERNATIONAL TRADE IN THE ERA OF NEO-GLOBALIZATION: DISINTEGRATION VS DIGITAL PARTNERSHIP. Res Econ Anthropol 42:7–13. https://doi.org/https://doi.org/10.1108/S0190-128120220000042001/FULL/XML
50. Parker S, Wu Z, Christofides PD (2023) Cybersecurity in process control, operations, and supply chain. Comput Chem Eng 171:108169. https://doi.org/https://doi.org/10.1016/J.COMPCHEMENG.2023.108169
51. Lejarza F, Kelley MT, Baldea M (2022) Feedback-Based Deterministic Optimization Is a Robust Approach for Supply Chain Management under Demand Uncertainty. Ind Eng Chem Res 61:12153–12168. https://doi.org/https://doi.org/10.1021/ACS.IECR.2C00099
52. Wren B (2022) Sustainable supply chain management in the fast fashion Industry: A comparative study of current efforts and best practices to address the climate crisis. Cleaner Logistics and Supply Chain 4:100032. https://doi.org/https://doi.org/10.1016/J.CLSCN.2022.100032
53. Iftikhar A, Ali I, Arslan A, Tarba S (2022) Digital Innovation, Data Analytics, and Supply Chain Resiliency: A Bibliometric-based Systematic Literature Review. Annals of Operations Research 2022 1–24. https://doi.org/https://doi.org/10.1007/S10479-022-04765-6
54. Orozonova A, Gapurbaeva S, Kydykov A, et al (2022) Application of smart logistics technologies in the organization of multimodal cargo delivery. Transportation Research Procedia 63:1192–1198. https://doi.org/https://doi.org/10.1016/J.TRPRO.2022.06.124
55. Klein M, Gutowska E, Gutowski P (2022) Innovations in the T&L (Transport and Logistics) Sector during the COVID-19 Pandemic in Sweden, Germany and Poland. Sustainability 2022, Vol 14, Page 3323 14:3323. https://doi.org/10.3390/SU14063323
56. Chen SH (2020) Digital humanities and the digital economy. Human-Machine Shared Contexts 359–383. https://doi.org/10.1016/B978-0-12-820543-3.00018-3
57. Pasiely MH, Pierce KE (2022) Large Scale Supply Chain Innovation in Canadian Wineries. International Journal of Data Science and Advanced Analytics 4:165–170
58. Chatzoudes D, Chatzoglou P (2022) Antecedents and effects of green supply chain management (GSCM) practices. Benchmarking ahead-of-print: https://doi.org/https://doi.org/10.1108/BIJ-09-2021-0524/FULL/XML
59. Cohen M, Cui S, Doetsch S, et al (2022) Bespoke supply-chain resilience: The gap between theory and practice. Journal of Operations Management 68:515–531. https://doi.org/https://doi.org/10.1002/JOOM.1184
60. Henrich J, Li JD, Mazuera C, Perez F (2022) Future-proofing the supply chain With supply chains in the spotlight, three new long-term transformation priorities form a fresh focus for competitive advantage
61. Kroes JR, Manikas AS, Gattiker TF, Castel MJ (2021) The operational impacts of Chief Supply Chain Officers in manufacturing firms. https://doi.org/10.1080/09537287.2021.1877840 33:1465–1480. https://doi.org/10.1080/09537287.2021.1877840
62. Elmghaamez IK, Attah-Boakye R, Adams K, Agyemang J (2022) The diffusion of innovation theory and the effects of IFRS adoption by multinational corporations on capital market performance: a cross-countryanalysis. Thunderbird International Business Review 64:81–108. https://doi.org/https://doi.org/10.1002/TIE.22244

63. Holmström Lind C, Kang O, Ljung A, Rosenbaum P (2022) Involvement of multinational corporations in social innovation: Exploring an emerging phenomenon. J Bus Res 151:207–221. https://doi.org/https://doi.org/10.1016/J.JBUSRES.2022.07.003
64. Cerar J, Decreton B, Nell PC (2023) What's in a Name? How Senior Managers use Name-Based Heuristics to Allocate Financial Resources in Multinational Corporations. Journal of Management Studies 60:1147–1177. https://doi.org/https://doi.org/10.1111/JOMS.12842
65. Jaiswal A, Arun CJ, Varma A (2021) Rebooting employees: upskilling for artificial intelligence in multinational corporations. https://doi.org/10.1080/09585192.2021.1891114 33:1179–1208. https://doi.org/10.1080/09585192.2021.1891114

Chapter 8
ESG and Circular Economy

Abstract This chapter focuses on the importance of Environmental, Social, and Governance (ESG) strategies in the context of sustainability. It begins with an introduction to the concept of ESG and its 3Ps—Purpose, Prosperity, and Preservation. The chapter then delves into the pitfalls of misguided sustainability projections, including overlooking the whole lifecycle, contamination, and greenwashing. Next, the chapter covers sustainable and green reporting, highlighting various frameworks such as the Global Reporting Initiative, Task Force on Climate-related Financial Disclosures, Sustainability Accounting Standards Board, Carbon Disclosure Project, and International Integrated Reporting Council. The chapter also discusses situational planning and investment management, outlining investor and institutional funds act, and the Freshfields Report. Finally, the chapter presents ESG case studies, including IBM. Overall, the chapter emphasizes the need for companies to incorporate ESG and circular economy strategies into their operations and investments for long-term sustainability.

Keywords Situational planning · Investment management · Sustainable reporting · CO_2 emission

8.1 Introduction to ESG and Its Strategies

Businesses across the world have started implementing a set of guidelines known as ESG, which stands for Environmental, Social, and Governance. In essence, these three aspects offer a framework for assessing a company's good benefits to the environment and society for investors and other stakeholders. ESG aspects are increasingly being published alongside financial performance, demonstrating the rising significance of ESG in the process of choosing investments [1, 2]. The ESG framework covers a wide variety of topics, such as risks and opportunities associated to climate change, the ecological effect of businesses, diversity and equity, consumer protection, human rights, and employee health and safety. It also takes into account matters like executive remuneration, diversity in leadership and the board, employee

S. Ramakrishna and B. Ramasubramanian, *Handbook of Materials Circular Economy*,
https://doi.org/10.1007/978-981-97-0589-4_8

and labor relations, and company ethics. A rising understanding of the effects that corporations have on the environment, society, and governance is reflected in the incorporation of ESG factors into investment decision-making. Investors may make better educated choices that are consistent with their beliefs and long-term financial objectives by taking these aspects into account besides financial success. ESG reporting and evaluation are therefore becoming more crucial for businesses looking to draw investment and prove their dedication to sustainability and moral behaviour [3].

The Three Levels of Strategy (Fig. 8.1) are three separate levels of strategy that affect the direction and success of a business. They were developed by famous management gurus Gerry Johnson and Kevan Scholes, namely, corporate supply chain strategy, business or tactical strategy, and functional or operational strategy. By eliminating inefficiencies in the supply chain and other organizational processes, a well-coordinated and harmonized strategy at each of these levels could contribute to the overall success of a business. The greatest level of strategic planning usually involves careful consideration and is entwined with the organization's goal and values [4].

Corporate supply chain strategy, business or tactical strategy, and functional or operational strategy are distinct, yet interrelated strategies used by companies to achieve their goals and objectives. These strategies are essential for ensuring the smooth functioning of a company. The corporate supply chain strategy outlines the overarching direction and structure of a company's supply chain. It involves making decisions about raw material sourcing, supplier management, distribution channel design, and logistics and transportation integration [4]. The corporate supply chain strategy establishes the framework for how the company will operate and sets the tone for the entire supply chain. The functional or operational strategy is the most detailed level of strategy and concentrates on specific functions within the company, such as manufacturing, marketing, finance, or human resources. These strategies support the

Fig. 8.1 Gerry Johnson and Kevan Scholes strategies

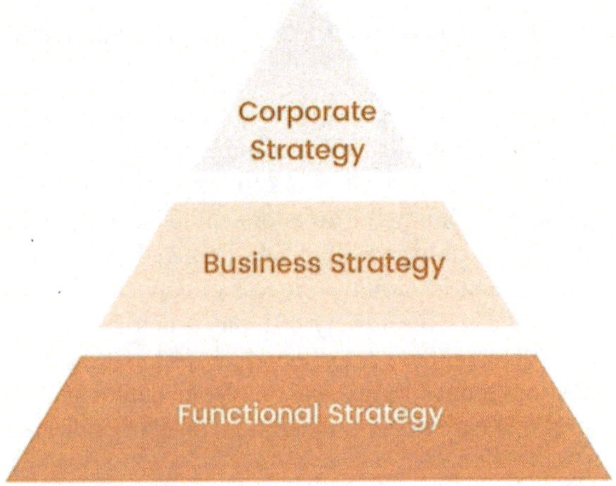

overall business and supply chain strategies and ensure that the company's resources are used effectively and efficiently. In the context of circular economy, these three levels of strategy can work together to promote a more sustainable and circular business model. The corporate supply chain strategy can include circular principles by reducing waste and increasing resource efficiency throughout the supply chain. The business strategy can prioritize circular products or services and target markets that value sustainable and circular practices. The functional or operational strategies can support these higher-level strategies by implementing circular practices within specific functions, such as designing products for circularity or reducing waste in manufacturing processes [4].

8.1.1 Economic Sustainability 3Ps—Purpose, Prosperity, and Preservation

Profit was given top priority in old company practices, with the social and environmental impacts of their activities being completely ignored. This strategy was motivated by a limited definition of success that solely considered immediate cash rewards. Modern companies understand that they must make a contribution to the greater good since the pursuit of profit alone is not sustainable. As purpose, prosperity, and preservation are interconnected, companies are now concentrating on addressing these dimensions. A company's dedication to having a beneficial influence on society is represented through its purpose. In other words, businesses understand that the communities in which they operate as well as their bottom line are impacted by what they do. Businesses with a purpose place a high value on ethical behavior, social responsibility, and the environment. Such procedures can eventually result in long-term profitability by fostering trust with stakeholders including clients, investors, and regulators [5].

The development of economic value for all parties involved, including consumers, investors, and the firm itself, is referred to as prosperity. In ensuring that their operations are profitable, businesses must also take into account the social and environmental consequences of their decisions. Creating shared value, where businesses cooperate to provide social and environmental advantages that also contribute to their financial success, is the key to reaching prosperity. Preservation highlights the significance of a business's environmental awareness and dedication to reducing its environmental effect. Businesses are taking initiatives to lessen their influence on the environment as they become more conscious of the possible effects of their actions. Businesses may adopt sustainable practices by decreasing waste, using less energy, and investing in renewable energy sources. Businesses must take a comprehensive strategy that balances Profit, People, and Planet in order to attain the three Ps. Instead of emphasizing one bottom line above another, this strategy calls on businesses to produce shared value for all stakeholders [5, 6]. It can be difficult for firms to strike this balance since they must choose between goals that may be at odds with one another. As a result, rather than concentrating exclusively on profit maximization, corporate executives must combine choices that consider the possible consequences on all three sectors [6].

Economic sustainability is concerned with maintaining and producing revenue from capital stock. This encompasses the notions of generating economic development and maximizing profits from a specific set of capital resources. The intent is to guarantee that subsequent generations have access to more capital goods per person than we have. Population increase hinders sustainable development since it reduces the available capital pool. Yet technical development can boost the wealth produced by a certain capital stock. According to the Triple Bottom Line Theory, capital can be in the form of natural, constructed, human, social, and financial assets [6].

Companies that have a deeper knowledge of capital can help promote sustainable growth. In order to do this, one sort of wealth must increase while another one declines. In order to offset the loss of non-renewable capital during the course of a project or firm, social investment can boost renewable capital. It's critical to share environmental costs and economic rewards fairly while also maintaining or raising the amount of capital assets. Fair distribution of renewable capital is challenging and out of the control of any one company at the local, national, and global levels. While though businesses are frequently held accountable for the uneven distribution of financial gains at the municipal and federal levels, the host government is mostly to blame [5, 6].

Profitability and the social structure are intertwined, which calls for tackling poverty via community development, giving host communities access to job and business possibilities, and considering local perspectives when making economic decisions. Obtaining raw materials for production and exploiting the environment as the ultimate waste sink are also involved. Economic sustainability ultimately aims to maximize one sort of wealth while limiting the loss of other types of value. Businesses should use the abatement cascade of Avoid-Eliminate-Help fix-Problem-solving methods when developing ways to lessen the decrease in any kind of assets [7].

8.1.2 Insights to Triple Bottom Line Theory

The Triple Bottom Line (TBL) Theory is a sustainable development strategy that takes into account social and environmental considerations in addition to financial success. According to the notion, businesses should strike a balance between their economic goals and how they will affect society and the environment in order to achieve sustainable growth. The TBL theory is based on three postulates: environmental sustainability, social sustainability, and economic sustainability. Economic sustainability is the capacity of a company to provide long-term economic value while balancing financial benefits with social and environmental factors. The ability of a company to enhance the wellbeing of its customers, staff, and communities is referred to as social sustainability [8, 9]. Sustainability in terms of the environment refers to a company's capacity to reduce its adverse effects on the environment. The TBL theory and the idea of environmental, social, and governance (ESG) are closely connected. Investors utilize ESG elements (Fig. 8.2), such as a company's effect on the environment, social issues, and corporate governance, to assess its non-financial performance. ESG factors play a significant role in determining a company's long-term viability and can impact its financial success [10].

The TBL hypothesis is connected to the circular economy idea as well. The circular economy promotes a closed-loop system where resources are utilized, reused, and recycled in an effort to reduce waste and increase resource efficiency [10]. This strategy is in line with the TBL theory's environmental sustainability postulate since it encourages resource efficiency and waste minimization, which lessens the influence that enterprises have on the environment. Through generating new business ventures and employment possibilities, as well as lowering reliance on resources and boosting social fairness, the circular economy may also contribute to the economic and social sustainability of society [10].

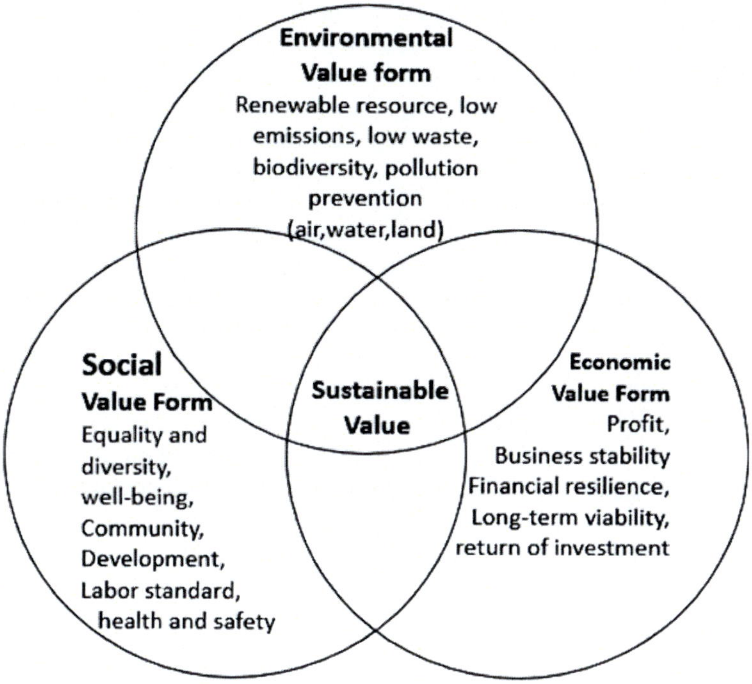

Fig. 8.2 The triple bottom line concept to evaluate organizations based on three dimensions: economic, social, and environmental [11]

8.2 Counting the Cost of Misguided Sustainability Projections

Businesses may profit greatly from sustainability programs, but it's crucial to avoid making assumptions that could have unforeseen repercussions. Sustainability efforts can have detrimental financial, environmental, and social effects when poorly planned or implemented, which can be costly to corporations and communities. Making sure the expenditures of these activities are in line with their financial objectives is one of the biggest issues facing businesses that are seeking sustainability [12]. Although using renewable energy sources could seem like a cheap approach to cut carbon emissions, these projects might have considerable expenses and aren't necessarily in line with a company's financial objectives. Over-reliance on renewable energy sources without sufficient planning and thorough examination of their true costs is one prevalent error. Businesses may make erroneous predictions regarding the costs and advantages of renewable energy sources, which can result in severe financial losses when the real expenses of putting these projects into action turn out to be higher than anticipated. Also, businesses might not consider the expense of incorporating renewable energy sources into their current infrastructure. It is necessary to make extra equipment, storage, and transmission infrastructure expenditures to integrate renewable energy sources, which may be expensive and time-consuming. Moreover, businesses could not take into account the long-term expenses of renewable energy sources. The long-term running costs of renewable energy sources may be lower than conventional energy sources, but their starting prices and estimated lifetimes may be higher. As a result, businesses who do not plan and budget for the complete lifetime of renewable energy sources risk suffering large financial losses [13].

In addition to reducing carbon emissions, sustainable production should also strive to minimize waste and promote the use of recycled materials. However, it is important to ensure that the production of recycled materials does not exceed the production of new materials. This is because the production of recycled materials requires energy and resources, which can result in negative environmental impacts. The circular economy strategy is frequently viewed as a means of striking a balance between environmental sustainability and economic growth. Yet, businesses who put profit ahead sustainability may embrace a circular economy strategy largely as a marketing tactic without making meaningful adjustments to their supply chain or production methods [8, 14]. This strategy may lead to company models that are not sustainable and a lack of actual advancement toward a circular economy. Businesses that put profit above sustainability in initial stages, may not put enough effort into developing a fully circular economy. This could provide the false impression that the company is not actively pursuing a sustainable future. Alternatively, businesses may employ the circular economy idea as a market research strategy to win and retain people who care about the environment, instead of focusing on profits in initial stages. For example, a firm can advertise a product as recyclable without making the necessary investments in building the infrastructure to collect and recycle the waste. This method of implementing priorities to profits than circular economy may lead to unsustainable business models that may not produce the intended environmental results [12, 13].

To express it in simple terms, embracing a circular economy approach should not only be seen as a useful marketing technique but also as a solution to the current environmental catastrophe. It's past time for businesses to understand that, in the long run, choosing sustainability instead of profit in the shorter term will result in a win–win situation and higher return in longer run. The advantages of investing in recycling expenses result in greater earnings over time. Prioritizing sustainability over profit is not only the right thing to do but also the smart thing to do [13, 14].

8.2.1 Overlooking the Whole Lifecycle

Sustainability is a multifaceted and complex notion that goes beyond only minimizing the negative effects of manufacturing on the environment. Also, it entails reducing a product's harmful effects over the course of its full existence. The latter can cause considerable environmental harm and dangers at both the micro and macro levels if it is not taken into account. Products that are advertised as sustainable yet have detrimental environmental effects during use and disposal can hurt people and communities on a micro level. For example, a product composed of eco-friendly materials that is not constructed to be easily repairable or recyclable may wind up in landfills, which would harm the environment. For those who live close to the garbage dump, this may result in respiratory issues, poisoned water supplies, and other detrimental health effects. The negative effects of unhealthy product lifecycles can, on a large scale, lead to problems with pollution, climate change, and resource depletion. For instance, items that are not made to be durable and easily recyclable may need to be replaced frequently, increasing resource use and waste production. Firms should focus on developing durable goods that take into account the full product lifetime, including the use of eco-friendly materials, production techniques, durability, reparability, and recyclability. Companies should also make investments in cutting-edge technology that enable closed-loop systems, which allow for the longest possible usage of the materials and goods. This strategy can save natural resources while lowering waste and pollution. Attaining sustainable product lifecycles necessitates a team effort from all stakeholders, from individual customers to big businesses [15, 16].

8.2.2 Contamination: Obstacles to Efficient Recycling

When non-recyclable elements are present in waste streams, recycling confronts tough obstacles. This not only taints the recycling process but also degrades the quality of the recycled goods, limiting their interest to consumers and reducing the demand for recycled goods. Additionally, organic trash like food scraps poses further difficulties since it contains germs and other organisms that might contaminate other recyclables. These impurities make recycling more challenging and potentially endanger the health of employees involved.

It's crucial to distinguish between fiction and fact in recycling. The idea that all kinds of paper waste may be recycled is among the most widespread ones. Paper towels and tissues with shorter fibers and colorants are hard to be recycled, despite the fact that many other forms of paper can. These items can contaminate the recycling process when they are placed in recycling bins, making it more challenging to effectively recycle other items. The possibility of deliberate contamination in recycling is another problem. Certain stakeholders, companies, or clients may attempt to sneak non-recyclable items into the recycling stream in an effort to save costs associated with disposal or simplify the procedure. Yet, this is a significant issue that might result in inefficient recycling, higher prices, and harm to the environment. Because tissues are often constructed of shorter fibers and have previously been used for their intended function, paper towels are frequently excluded from recycling because they might be more challenging to recycle. Moreover, they may contain food particles, cleaning agents, or other contaminants that lower the quality of the recovered material. Although while it is technically feasible to recycle some paper towels and tissues, such as those manufactured from 100% recycled fibers and devoid of any pollutants, it is not a generally practiced method in the majority of recycling systems. Paper towels and tissues should generally advisable to be made out of biodegradable materials.

8.3 Sustainable and Green Reporting

The process of informing stakeholders on a company's environmental, social, and governance (ESG) performance is known as sustainable reporting. A company's sustainability goals, strategies, and advancement are covered in depth in a sustainability report, which is normally where this information is published. As stakeholders demand greater openness and responsibility from businesses, sustainable reporting—which can be voluntary or required by law is growing in popularity. The most common reporting standards and associated values are covered in this section. In the following section, we will analyze the feasible reporting standards offered by the Global Reporting Initiative (GRI), Sustainability Accounting Standards Board (SASB), Task Force on Climate-related Financial Disclosures (TCFD), Carbon Disclosure Project (CDP), and International Integrated Reporting Council (IIRC) [5, 10].

Founded year	Founded organisation	Applicability	Purpose	Where to report	Sector focus	Flexibility	ESG coverage	Website
1997	Global reporting initiative (GRI)	Global	Sustainability reporting standards	GRI sustainability reporting standards	All sectors	Highly flexible	All ESG	https://www.globalreporting.org/
2011	Sustainability accounting Standards Board (SASB)	US	Industry-specific sustainability disclosure standards	SASB standards	Industry-specific	Less flexible	All ESG	https://www.sasb.org/
2015	Task force on climate-related financial disclosures (TCFD)	Global	Climate-related financial disclosure recommendations	TCFD recommendations	All sectors	Highly flexible	E and C	https://www.fsb-tcfd.org/
2000	Carbon disclosure project (CDP)	Global	Climate, water, and forest risk disclosure	CDP Climate Change, Water Security, and Forests Questionnaires	All sectors	Moderately flexible	E and C	https://www.cdp.net/
2013	International integrated reporting council (IIRC)	Global	Integrated reporting framework	Integrated report	All sectors	Moderately flexible	All ESG	https://integratedreporting.org/

8.3.1 Global Reporting Initiative

The United Nations Environment Programme (UNEP) and the Coalition for Environmentally Responsible Economies (CERES) founded the Global Reporting Initiative (GRI) in 1997. The GRI is an independent, global organization that created a framework for sustainability reporting that is currently utilized by thousands of businesses and organizations across more than 100 nations. The goal of the GRI framework is to give businesses a consistent language to report on their sustainability performance and to make it possible for stakeholders to compare performance across various enterprises [17]. The GRI Standards, which provide a comprehensive set of indicators for reporting on a wide variety of topics, serve as the framework's cornerstone. The GRI framework is applicable to all organizations, regardless of size, sector, or location. While the GRI Standards are not legally binding, many organizations choose to follow them voluntarily, in order to demonstrate their commitment to sustainability and to meet stakeholder expectations. Organizations are advised to begin with GRI 1: Foundation 2021, which explains the fundamental principles and guidelines of GRI Standards for sustainability reporting [17]. GRI 2: General Disclosures 2021 contains information on reporting procedures and organizational details, allowing for a comprehensive assessment of an organization's impacts. GRI 3: Material Topics 2021 provides guidance on recognizing important topics and reporting the process of managing them (Fig. 8.3). Here we focus on GRI 3, as the book is relative to materials circularity, however the standard reporting procedures can be found at GRI—Home (globalreporting.org).

Organizations must go by a set of requirements set out by the Global Reporting Initiative (GRI) when reporting on their sustainability activities. The first requirement mandates that organizations follow the GRI's reporting guidelines. Second, companies must submit the disclosures defined in GRI 2: General Disclosures 2021, which includes details on the governance, operations, and policies of the company. Organizations are required by Requirement 3 to specify their material themes, as described in GRI 3: Material Topics 2021. Organizations are then required to report on the disclosures in GRI 3 that pertain to their material issues under Requirement 4 [18]. Organizations must publish data from the GRI Topic Standards for each of their material themes in order to meet Requirement 5. Moreover, Requirement 6 requires companies to justify any disclosure omissions or requirements that they are unable to meet. Organizations are required to provide a GRI content index, indicating which disclosures have been recorded and where they may be located, in accordance with Requirement 7. Organizations are also required by Requirement 8 to submit a statement of use that specifies which GRI Standards they have used in their reporting. Last but not least, Requirement 9 stipulates a number of specific disclosures that organizations must include in their reports. These disclosures include organizational information, entities included in sustainability reporting, reporting period, frequency, and contact information, restatement of information, external assurance, process to identify material topics, and a list of material topics. Organizations may make sure that their sustainability reports adhere to the GRI Guidelines and give a thorough

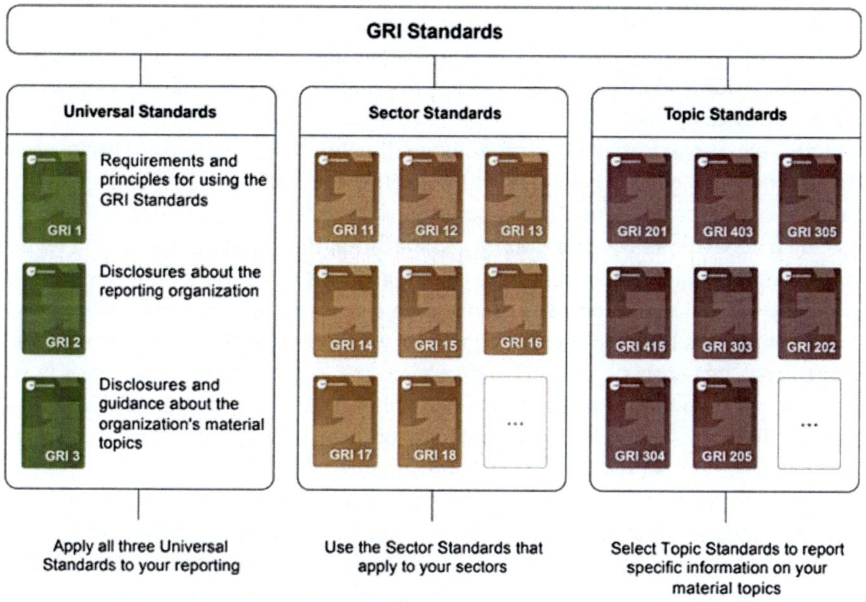

Fig. 8.3 GRI standards used for ESG reporting

overview of their sustainability actions by adhering to these responsibilities. The sectors involved in GRI reporting can include any industry or sector, as sustainability impacts are relevant to all organizations. However, some sectors may be more likely to report using the GRI Standards, such as those in the energy, mining, manufacturing, and financial industries, where ESG issues are particularly relevant [18].

8.3.2 Task Force on Climate-Related Financial Disclosures

The Task Force on Climate-related Financial Disclosures (TCFD) was established by the Financial Stability Board (FSB) to provide a voluntary, consistent framework for companies to disclose climate-related financial risks and opportunities. The TCFD framework is intended to help companies better understand and communicate their exposure to climate-related risks and opportunities and to help investors and other stakeholders make more informed decisions. In compliance with the TCFD recommendations, it is essential for companies to disclose information regarding the potential financial impacts of climate change on their business operations. TCFD's 11 recommendations provide a framework for companies to assess and disclose climate-related risks and opportunities. This disclosure includes identifying the potential physical, transition, and liability risks, as well as assessing the resilience of the company's strategy to a range of climate scenarios. By disclosing this information, companies can better inform investors and stakeholders of the potential impacts of climate change on their business and demonstrate their commitment to managing these risks. Ultimately, this leads to better decision-making and more sustainable business practices [17].

Within the context of the Task Force on Climate-related Financial Disclosures (TCFD), risk management entails identifying and evaluating potential financial risks related to climate change, including physical risks (such as infrastructure damage resulting from extreme weather events) and transition risks (such as policy and regulatory changes impacting the value of fossil fuel assets). On the other hand, opportunities refer to the financial benefits that companies can potentially realize by addressing climate change, such as cost savings from energy efficiency measures or new revenue streams from low-carbon products and services. Income statements and cash flow statements are two financial statements that are typically employed in TCFD analysis [17]. Income statements provide details on a company's revenues, expenses, and profits, while cash flow statements show the movement of cash in and out of a company. Both of these statements can be leveraged to analyze the financial effects of climate change on a company, such as variations in revenue and expenses due to climate-related incidents or investments in low-carbon technologies. The advantages of TCFD are significant. By offering greater transparency and consistency in climate-related financial disclosures, TCFD can assist investors in making better-informed decisions and encourage companies to invest in low-carbon technologies and approaches [17].

8.3.3 Sustainability Accounting Standards Board

A non-profit organization called the Sustainability Accounting Standards Board (SASB) was founded with the purpose of creating and promoting sustainability accounting standards that will assist publicly traded companies in telling investors important ESG information. The SASB standards are intended to be market-informed, evidence-based, and sector-specific, which means they take into account the sustainability challenges that are most likely to have an impact on a certain industry. In order to comply with SASB reporting guidelines, a corporation must provide significant ESG data in its annual reports, and sustainability reports. Investors should be able to evaluate ESG performance across businesses in the same industry easily thanks to the information's presentation in a structured style that adheres to SASB guidelines. Almost 75 industries and 11 sectors are covered by SASB standards, including, among others, transportation, energy, financial services, and agriculture. Each sector has its own set of standards, which are created via a thorough process that includes peer review, stakeholder participation, and materiality analysis. The disclosure themes covered by the standards often cover, among other things, the industry's effects on the environment, labor practices, product quality and safety, data privacy and security, and corporate governance. Depending on a company's sector and business model, SASB standards may or may not apply [19, 20].

The specific metrics and targets utilized for measuring sustainability depend on the industry and the particular environmental concern being addressed. For example, energy consumption and greenhouse gas emissions can be assessed by calculating the total energy consumed, energy intensity per unit of output, and total greenhouse gas emissions. Targets may include achieving a specific reduction in energy usage and emissions over a specified timeframe. Water management can be measured by tracking total water withdrawals, water intensity per unit of output, and the proportion of recycled or reused water. Targets may include reducing water withdrawals, improving water efficiency, and achieving water neutrality. Regarding human capital management, metrics such as employee turnover, diversity, and training hours can be used. Targets may involve increasing employee retention, promoting diversity and inclusion, and investing in employee development. Further, supply chain management can be assessed by measuring the percentage of suppliers audited for environmental and social performance, the percentage of suppliers meeting sustainability criteria, and the percentage of purchases from local suppliers. Targets may aim to increase the percentage of sustainable suppliers, decrease supply chain risks, and improve supplier performance. Lastly, product quality and safety can be evaluated by tracking the percentage of products tested for safety and quality, the number of product recalls, and the percentage of products meeting sustainability standards. Targets may aim to improve product safety and quality, reduce product recalls, and increase the percentage of sustainable products [17, 21, 22].

8.3.4 Carbon Disclosure Project (CDP)

The Carbon Disclosure Project (CDP) is a non-profit organization that collaborates with organizations, cities, and states to disclose their environmental impact, especially in terms of greenhouse gas emissions. The CDP gathers and assesses data on the environmental performance of these entities and shares the information with investors, policymakers, and the public. To utilize the CDP, organizations should follow these steps: identify their sector, choose the appropriate questionnaire, respond to the questionnaire with accurate and complete responses, submit the questionnaire, and use the analysis and feedback received from the CDP to make changes and report progress over time. The CDP assesses environmental performance in categories such as Climate Change, Water Security, Forests, and Supply Chain. The CDP evaluates organizations' environmental impact using metrics such as carbon emissions, energy use, water use, and supply chain sustainability, including supplier sustainability performance and supply chain risks [23, 24].

Carbon tax is a policy tool used by governments to incentivize companies and individuals to reduce their carbon footprint by making it more expensive to emit greenhouse gases. The tax is typically levied on the burning of fossil fuels such as coal, oil, and gas, which are significant sources of carbon emissions [25, 26]. As of 2021, there are over 70 carbon pricing initiatives in place worldwide, including carbon taxes, emissions trading schemes, and other market-based mechanisms. Sweden implemented the world's first carbon tax in 1991, which is set at SEK 1190 (around $138) per metric tonne of carbon dioxide. Switzerland introduced a carbon tax on heating oil and natural gas in 2008, currently at CHF 96 (around $104) per metric tonne of carbon dioxide. Canada implemented a federal carbon tax in 2019, set at CAD 20 (around $16) per metric tonne of carbon dioxide, which is planned to increase to CAD 170 (around $136) by 2030. In 2013, the United Kingdom introduced a carbon floor price of £18.08 (around $25) per metric tonne of carbon dioxide, and the price is planned to increase to £30 (around $41) by 2022 [27, 28]. One of the challenges of implementing a carbon tax is the potential impact on low-income households and industries that heavily rely on fossil fuels. To address this issue, some governments have introduced measures such as rebates or exemptions for low-income households or energy-intensive industries. Additionally, some governments have used revenue generated from the carbon tax to fund investments in renewable energy, energy efficiency, or other green initiatives. Another challenge is the possibility of carbon leakage, where companies move their operations to countries with weaker climate policies, leading to no net reduction in emissions [29, 30].

8.3.5 International Integrated Reporting Council

The International Integrated Reporting Council (IIRC) is a global non-profit organization that advocates for integrated reporting to promote transparency and sustainable outcomes for companies, investors, and other stakeholders. Integrated reporting involves the integration of financial, environmental, social, and governance (ESG) performance to provide a holistic view of an organization's value creation story. The IIRC provides guidance and resources to help organizations develop integrated reports that offer insight into the links between financial and non-financial factors, risks and opportunities, and the organization's strategy and governance. The organization believes that integrated reporting can lead to better decision-making, capital allocation, and sustainability outcomes [31, 32]. To utilize integrated reporting, organizations should identify their value creation story, engage with stakeholders, integrate sustainability into business strategy, and continuously improve their reporting and performance over time [33, 34]. The IIRC's framework for integrated reporting includes organizational overview, business model, risks and opportunities, performance, and outlook. Performance across financial, ESG factors, the IIRC uses a range of metrics and analysis, such as traditional financial metrics (e.g., revenue, profit, and return on investment), environmental metrics (e.g., greenhouse gas emissions, water use, and waste management), social metrics (e.g., employee engagement, customer satisfaction, and community relations), and governance metrics (e.g., board composition, executive compensation, and risk management) [31].

8.4 Situational Planning and Investment Management

In the context of ESG, emergency and situational planning typically involves preparing for natural disasters, environmental incidents, and social disruptions, as well as other events that may impact corporate governance. This can include developing plans for crisis communication, stakeholder engagement, and business continuity, as well as identifying and assessing potential risks and hazards associated with ESG issues [35].

The development of a Crisis Management Plan (CMP) and Crisis Management Team (CMT) necessitates a thorough comprehension of the significant factors that differentiate crisis management from routine operational management. Crisis management refers to the process of preparing for, managing, and recovering from a crisis that jeopardizes the organization's ability to operate effectively. The factors that set crisis management apart from routine operational management include several aspects. Firstly, crisis management demands a centralized leadership-based command and control structure, with a designated leader possessing the authority to make critical decisions swiftly. Secondly, decisions must be made quickly, and tasks must be executed rapidly in a crisis situation to minimize the damage caused by the crisis. Thirdly, during a crisis, the information available to decision-makers is often incomplete or inaccurate, and leaders must make decisions based on limited information. Fourthly, the responsibilities and tasks assigned to individuals and teams during a crisis may differ significantly from their usual roles and responsibilities [36].

Furthermore, the consequences of mismanagement during a crisis can be severe, including financial losses, reputational damage, and legal liabilities. Crises can significantly impact an organization's stakeholders, including employees, customers, suppliers, and the broader community. Moreover, crises often attract media attention, which can further complicate crisis management. Additionally, in some cases, executives may face criminal prosecution if their actions during a crisis are found to be negligent or unlawful. Finally, crisis management can be a highly stressful and exhausting process, requiring individuals and teams to work long hours under intense pressure. Despite the inevitability of crises, many companies perform poorly in managing them. According to Pricewaterhouse Coopers (PwC), nearly two-thirds of companies they surveyed had experienced at least one corporate crisis in the 5 years preceding the survey, with an average of three crises [37]. A crisis management model is a conceptual framework used to describe how crises develop and can be managed. Scenario-based crisis preparedness is becoming increasingly ineffective due to operational complexity and unpredictable crisis events, making a more robust and flexible approach to crisis management necessary [35].

A simple and uncomplicated goal, financial benefit for shareholders is prioritized by certain investment managers. ESG issues, for example, might add complexity when larger investing objectives are taken into account. ESG investing proponents argue that these variables imply significant risks that standard financial analysis may not adequately account for. They assert that carefully concentrating on the financial interests of beneficiaries requires the incorporation of ESG concerns [37]. Fiduciaries have a responsibility to act in the beneficiaries' best interests and carry out their duties responsibly. The fiduciary's principal duty is to operate in the beneficiary's best interests while abstaining from conflicts of interest or self-dealing. This can become more complicated when ESG considerations are involved. The default position is normally to maximize financial return if no instruction is provided by the original documents or beneficiaries. The second responsibility is to treat beneficiaries impartially, which is essential for pooled funds with diverse interests but not difficult for funds with comparable goals. The third requirement is to exercise care, which entails carefully considering decisions and making wise financial judgments. Although there are some differences in these requirements for various types of fiduciaries, the obligations are generally similar [35, 37].

8.4.1 Investor Act

The Uniform Prudent Investor Act (UPIA) is a legal framework that governs the management of trust assets by trustees, replacing the outdated "Prudent Man" rule. Under the UPIA, trustees are held to a fiduciary duty to act in the best interests of beneficiaries and avoid any conflicts of interest or self-dealing. The UPIA requires trustees to manage trust assets as a prudent investor would, using reasonable care, skill, and caution. Trustees must diversify the portfolio according to Modern Portfolio Theory and exercise prudence when hiring skilled advisors, accountants, and lawyers while minimizing costs. The UPIA recognizes that asset classes considered to be prudent investments have evolved over time and does not prohibit any investment outright. Instead, trustees must make investment decisions based on the standard of a prudent investor [38, 39]. The UPIA was first introduced in the Commonwealth of Massachusetts in the early 1800s and has since been adopted by other states. Trustees must comply with the "The Restatement (Second) of Trusts: Prudent Man Rule (1959)," which stipulates that trustees can only make investments that a prudent person would make, with the goal of preserving the estate and generating regular income. Trustees must also follow any statutes governing investments by trustees and adhere to the terms of the trust [40]. The UPIA recognizes that the investment landscape has changed since the "Prudent Man" rule was established, and thus is more flexible and adaptable to various investment options. Trustees are required to use reasonable care and skill to preserve trust property and make it productive, unless the terms of the trust dictate otherwise. Overall, the UPIA ensures that trustees act in the best interests of beneficiaries and manage trust assets prudently [41].

8.4.2 Institutional Funds Act

Non-profit and charitable organizations are bound by state laws and the Uniform Prudent Management of Institutional Funds Act (UPMIFA), which updated a prior act that limited non-profits from expending funds if their capital fell below the initial value of donations. Fiduciaries managing charitable organizations must adhere to principles of loyalty, impartiality, and prudent investment, giving primary consideration to donor intent expressed in the gift instrument. UPMIFA authorizes institutions to allocate funds as they deem judicious, subject to donor instructions. Fiduciaries are required to act in the charity's best interest and may consider the organization's mission when making investment decisions. In 2016, the US Department of Labor proposed the "Fiduciary Rule" to broaden the definition of an "investment advice fiduciary," but in March 2018, the US Fifth Circuit Court of Appeals invalidated the rule. The Freshfields Report, commissioned by the Asset Management Working Group of the United Nations Environment Programme Finance Institute (UNEP FI) in 2005, examined fiduciary responsibility in several nations [42, 43].

8.4.3 Freshfields Report

The Freshfields Report, commissioned by the Asset Management Working Group of the United Nations Environment Programme Finance Initiative (UNEP FI) in 2005, explores the issue of fiduciary responsibility in several countries around the world. The report was motivated by the need for international guidance on fiduciary duties and investment practices for asset managers and other fiduciaries in the context of environmental, social, and governance (ESG) issues. One of the key needs for the Freshfields Report was the lack of clarity and guidance on how fiduciaries should take into account ESG factors in their investment decisions. While fiduciaries have a legal obligation to act in the best interests of their clients or beneficiaries, there was no clear consensus on how to incorporate ESG issues into the investment process. This led to inconsistencies in the way that different fiduciaries approached ESG considerations and a lack of transparency for clients and beneficiaries [44, 45].

However, the Freshfields Report faced several challenges in providing comprehensive guidance on fiduciary duties and ESG issues. One of the main challenges was the lack of uniformity in fiduciary law and investment practices across different countries and regions. The report had to consider the unique legal and regulatory frameworks of each country and how they interacted with ESG considerations. Another challenge was the need to balance fiduciary duties with ESG considerations. While fiduciaries have a responsibility to maximize returns for their clients or beneficiaries, they also have a duty to act in the best interests of society and the environment. The report had to address how fiduciaries could reconcile these sometimes-competing obligations. Additionally, the Freshfields Report had to address the practical challenges of incorporating ESG factors into investment decisions. This included identifying relevant ESG issues and data sources, evaluating the materiality of ESG factors, and determining the appropriate level of ESG integration for different types of investments.

Despite these challenges, the Freshfields Report provided valuable guidance on fiduciary responsibility and ESG issues for asset managers and other fiduciaries around the world. The report emphasized the importance of incorporating ESG considerations into investment decisions and provided practical guidance on how to do so while maintaining fiduciary duties [42, 45].

8.5 ESG Case Studies

Sustainability-related consequences of ESG are numerous. For instance, businesses may support the global effort to prevent climate change by concentrating on environmental aspects like lowering carbon emissions and controlling waste. Similar to how individuals and organizations may boost their reputations and forge closer ties with stakeholders by giving social aspects like community involvement and human rights top priority. Lastly, organizations may increase their credibility and confidence with investors by concentrating on governance elements like accountability and transparency.

8.5.1 Sustainability Indicators and Financial Stability in Russian Oil and Gas

Sustainability Performance and Financial Stability in the Russian Oil and Gas Industry. The Russian oil and gas industry holds significant importance for the nation's economy, yet it has encountered formidable challenges arising from global crises and international sanctions. Consequently, there has been an intensified emphasis on adopting sustainable practices within the industry. As early as 1996, the concept of sustainable development was embraced, albeit with limited effectiveness in shaping corporate policies. Subsequently, in 2002, the government demonstrated a commitment to sustainability through the introduction of the Environmental Doctrine and Strategy, which sparked modest advancements in non-financial reporting [46]. However, in recent times, economic and political upheavals have exerted adverse impacts on corporate social responsibility (CSR) efforts and overall sustainability development. The lingering influence of the post-communist environment, coupled with regulatory constraints, has further complicated the implementation of CSR initiatives. Hence, the urgent need to enhance social and environmental responsibility standards within the oil and gas sector has become a critical imperative for Russia's business community [46, 47].

Orazalin et al. used the GRI framework, it comprehensively assesses sustainability practices in the region [48]. Empirical findings support the notion that improved sustainability performance positively influences financial stability. The study also identifies firm-specific characteristics like financial capacity, leverage, size, and age as significant factors affecting financial stability. By exploring sustainability practices in the oil and gas industry and analysing economic, environmental, and social indicators through GRI4 standards, this research contributes valuable insights to the field. Improved sustainability performance is expected to enhance financial stability for top oil and gas companies in Russia (H1), with positive associations between economic (H1a), environmental (H1b), and social (H1c) sustainability performance and financial stability. This study explores sustainability practices of oil and gas companies in Russia and their impact on financial stability. Based on panel data analysis from 2012–2016, improved sustainability performance is found to enhance financial stability. Policymakers, regulators, and investors can benefit from these findings by encouraging companies to adopt GRI standards for more informative and transparent reporting. Limitations include the sample size and focus on oil and gas industry, suggesting future research on other industries and emerging markets.

8.5.2 *Factors in Climate Risk Disclosure by Brazilian Companies in Sustainability Reports*

The climate risk information disclosure in GRI reports of Brazilian-listed companies from 2009 to 2014 were explored by Kouloukoui et al. The sample consisted of 67 companies with a total of 402 observations. Results showed that although Brazilian companies disclosed some information on climate risks, the level of disclosure remained relatively low. The study found significant and positive relationships between climate risk disclosure and firm size, financial performance, and country of origin. Over the years, the number of companies disclosing climate risk information increased. The most disclosed term was emissions, followed by greenhouse gas and CO_2. However, terms related to natural disasters received little attention. The study suggests that both government regulation and creditor incentives could encourage greater disclosure of climate risks [49].

8.5.3 *IBM*

IBM engages with GRI, SASB, TCFD, Stakeholder Capitalism Metrics, and UN SDGs. Their IBM Sustainability Accelerator supports non-profits and vulnerable groups using hybrid cloud and AI, focusing on sustainable agriculture and clean energy with a $30M commitment. They prioritize energy efficiency in data centers, aiming for a Power Usage Effectiveness (PUE) value closer to 1. [50]

IBM's 2022 energy use decreased by 1.5% compared to 2021, with approximately 2,448,000 MWh consumed globally, 80% of which was electricity. They increased renewable electricity consumption to about 1,299,000 MWh, accounting for 65.9% of total electricity use (Fig. 8.4a–d). IBM aims for residual emissions to be 350,000 metric tons or less of CO_2-equivalent, covering Scope 1, Scope 2, and specific Scope 3 emissions associated with electricity consumption at co-location data centers.

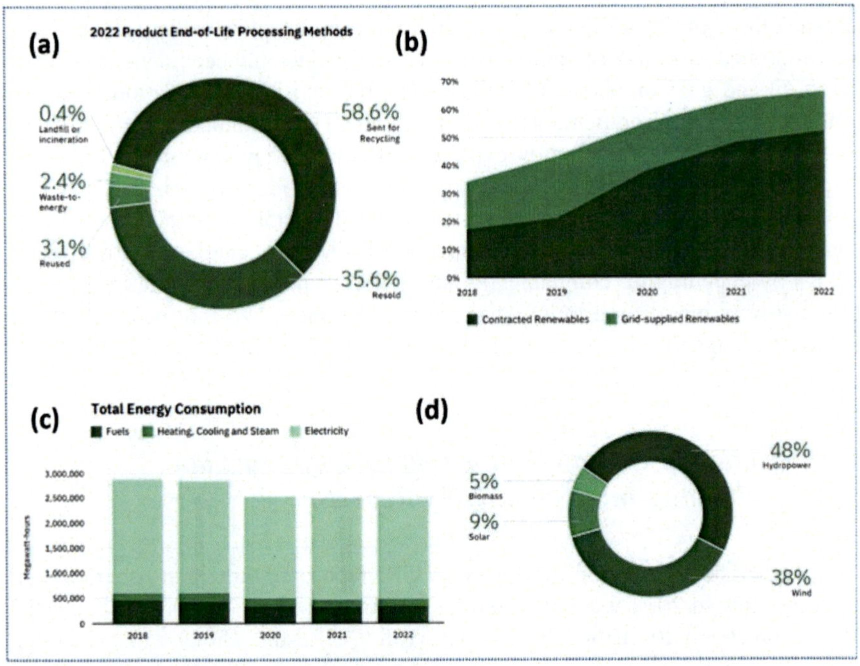

Fig. 8.4 **a** End of Life processing methods of used products from IBM, **b** Use of Renewable Electricity as Percent of Global Electricity Consumption, 2022, **c** Total energy consumption as per 2022 ESG files of IBM and **d** % of energy sources used for power generation at IBM in 2022 [50]

IBM's eco-friendly changes as shown in Fig. 8.5 include replacing nonessential polybags with paper envelopes, saving 1.7 MT of polyethylene bags yearly, switching zip-lock plastic bags with paper envelopes, saving 1 MT of zip-lock bags annually, replacing LDPE air pillow void filler with 100% recycled content LDPE air pads, saving 0.3 MT of virgin LDPE material, and using 30% recycled content PET banding and 100% recycled content LDPE stretch wrap, saving 0.3 MT of virgin polypropylene and 0.4 MT of virgin LDPE wrap per year, respectively.

8.5.4 Apple Inc

40% emissions decrease in Apple's entire value chain since 2015. 100% renewable energy sourced for all Apple facilities. 213 suppliers committed to 100% renewable electricity for Apple production. $4.7B issued in green bonds to reduce global emissions, and increased use of 100% recycled aluminium in product enclosures, reducing carbon emissions by 68% since 2015 (Fig. 8.6). Apple aims to reduce their carbon footprint, promote product recycling and sustainable material sourcing. Water conservation measures are also a priority. Apple focuses on responsible supply chain

(a)

Product Reuse and Recycling	2018	2019	2020	2021	2022
Total end-of-life products and product waste processed (metric tons x 1,000)	28.3	20.8	16.9	18.0	12.4
% By weight of total end-of-life products and product waste sent by IBM's product end-of-life operations to landfill or incineration for treatment (goal not to exceed a combined 3% by weight)	0.7	0.8	0.5	0.3	0.4

(b)

Energy and Climate Change	2018	2019	2020	2021	2022
Total operational energy consumption in megawatt-hours	2,870,000	2,837,000	2,529,000	2,486,000	2,448,000
Renewable electricity procurement as % of total electricity consumption (goal 75% by 2025, and 90% by 2030)*	33.6%	42.9%	55.6%	62.7%	65.9%
Operational GHG emissions reduction as % of 2010 base year (goal 65% by 2025)	26.6%	31.0%	51.5%	59.5%	63.3%
Total operational GHG emissions covered by our goal in metric tons of CO_2-equivalent	862,000	811,000	570,000	476,000	431,000

*Renewable electricity procurement includes contracted purchases and renewable electricity that automatically comes to IBM via routine grid power

(c)

Emissions Covered by IBM's Fifth-Generation Goal (Metric Tons of CO_2-equivalent)	2018	2019	2020	2021	2022
Scope 1 (direct emissions)*	105,000	100,000	74,000	79,000	79,000
Scope 2 market-based (indirect emissions)**	505,000	460,000	262,000	221,000	183,000
Scope 3 (indirect emissions)'	252,000	251,000	234,000	176,000	169,000
Total emissions covered by IBM's current goal	**862,000**	**811,000**	**570,000**	**476,000**	**431,000**
Reduction of GHG emissions against base year 2010	**26.6%**	**31.0%**	**51.5%**	**59.5%**	**63.3%**

*Emissions associated with IBM's use of fuels for building operations and transportation, as well as from the use of refrigerants and chemicals with a global warming potential

**Emissions from IBM's use of electricity, cooling, heat and steam at IBM-managed locations, accounting for our purchases of renewable electricity

'Includes only those emissions associated with the generation of electricity consumed by IBM's data centers located in third-party managed facilities

Fig. 8.5 **a** Reuse and recycling rates in IBM, **b** initiatives for climatic action by IBM and **c** GHG emissions (Records of ESG_IBM 2020) [50]

practices, supports employees through various initiatives, engages in community and educational programs, and emphasizes user privacy and data security within the broader ESG framework [51].

8.5.5 McKinsey ESG Approach

McKinsey's ESG approach involves integrating ESG considerations into clients' core business strategies and operations by identifying ESG risks and opportunities aligned with overall objectives. They assist in establishing meaningful ESG performance metrics for transparency and accountability. Engaging stakeholders is a priority, understanding their perspectives on ESG issues to build trust and find collaborative solutions. McKinsey helps companies develop tailored sustainability strategies and manage ESG-related risks for long-term resilience. They promote supply chain sustainability, support diversity, equity, and inclusion efforts, and advise on transitioning to cleaner energy sources amidst growing climate change concerns. Apart

Greenhouse gas emissions

		Fiscal year				
		2021	**2020**	**2019**	**2018**	**2017**
Corporate emissions[1] (metric tons CO_2e)	**Scope 1 (gross emissions)**	**55,200**	**47,430**	**52,730**	**57,440**	**47,050**
	Natural gas, diesel, propane	40,070	39,340	40,910	42,840	36,210
	Fleet vehicles	12,090	4,270	6,950	11,110	8,300
	Process emissions[2]	3,040	3,830	4,870	3,490	2,540
	Scope 2 (market-based)	**2,780**	**0**	**0**	**8,730**	**36,250**
	Electricity	0	0	0	8,730	36,250
	Steam, heating, and cooling[3]	2,780	-	-	-	-
	Scope 3 (gross emissions)[4]	**23,130,000**	**22,550,000**	**24,980,000**	**25,070,000**	**27,330,000**
	Business travel[5]	22,850	153,000	326,000	337,000	121,000
	Employee commute[6]	85,570	134,000	195,000	183,000	172,000
	Corporate carbon offsets[7]	-167,000	-70,000	-	-	-
Product life cycle emissions[8] (metric tons CO_2e)	Manufacturing (purchased goods and services)	16,200,000	16,100,000	18,900,000	18,500,000	21,100,000
	Product transportation (upstream and downstream)	1,750,000	1,800,000	1,400,000	1,300,000	1,200,000
	Product use (use of sold products)	4,990,000	4,300,000	4,100,000	4,700,000	4,700,000
	End-of-life treatment	80,000	60,000	60,000	50,000	40,000
	Product carbon offsets[9]	-500,000	-	-	-	-
Total gross carbon footprint (without offsets)[10] (metric tons CO_2e)		**23,200,000**	**22,600,000**	**25,100,000**	**25,200,000**	**27,500,000**
Total net carbon footprint (after applying offsets)[10] (metric tons CO_2e)		**22,530,000**	**22,530,000**	**25,100,000**	**25,200,000**	**27,500,000**

Fig. 8.6 GHG emissions of Apple Inc, ESG reports 2022 © Apple Inc. [51]

from this, McKinsey's ESG priorities, identified through materiality assessments, are at the core of their sustainable and inclusive growth strategy, driven by their commitment to responsible business practices. They prioritize sustainable growth, shaping how they operate, serve clients, and contribute to communities. To reduce carbon emissions, they are actively transitioning to electric vehicles, with EV-only policies covering over 50% of their global car fleet, leading to a threefold increase in hybrid and EV car usage since 2019, reaching 27% in 2022 from 9% before. McKinsey considered calculations for 2025, 2030, and 2050, with business travel accounting for about 80% of emissions in 2019. They used forecasted carbon prices from UNPRI's "Required Policy Scenario," weighting them based on their emissions across different regions. UNPRI predicts carbon prices of $45–85 per ton by 2030 and $87–175 per ton by 2050. However, these pricing regulations are not expected to significantly affect the firm's financial or strategic aspects since they don't operate in an emissions-intensive industry and have committed to achieving net zero by 2030 [52].

8.5.6 Others

Keppel Corporation Limited is a multinational company that focuses on offshore and marine, infrastructure, and property businesses. The company has made sustainability a key priority and is committed to achieving net-zero carbon emissions by 2050. To achieve this goal, Keppel has invested heavily in renewable energy and clean technologies, while also launching initiatives aimed at reducing waste and promoting circular economy principles. In 2020, the company was able to reduce its carbon emissions by 24%, and it has a target to achieve a 28% reduction by 2030 [53, 54].

Singtel, one of the largest telecommunications companies in Singapore, has also integrated sustainability into its corporate strategy. The company has set targets to reduce its carbon footprint and increase the use of renewable energy, with initiatives aimed at promoting digital inclusion and community well-being. Singtel has received recognition for its sustainability efforts from the Dow Jones Sustainability Index and the Global 100 Most Sustainable Corporations in the World. In 2021, the company was able to reduce its carbon emissions by 3.5% and increase its use of renewable energy to 10% [55].

Sembcorp Industries, a leading energy and environmental solutions provider, is committed to reducing its carbon footprint and increasing the use of renewable energy. The company has launched initiatives aimed at promoting sustainable water management and waste reduction. Sembcorp has also received recognition for its sustainability efforts from the Dow Jones Sustainability Index and the Carbon Disclosure Project. In 2020, the company was able to achieve a 5.6% reduction in carbon emissions and had 3.7GW of renewable energy capacity. One of them is Marina Bay Sands, an iconic integrated resort, that has set ambitious targets to reduce its carbon emissions intensity by 45% by 2030 [56].

8.6 Conclusion

In conclusion, achieving sustainable economic growth while reducing the effects of climate change requires the implementation of ESG methods and circular economy concepts. To avoid greenwashing, take into account a product's whole lifespan, and engage in sustainable methods and clean technology. For openness and accountability to be upheld, green and sustainable reporting is essential. Situational planning and investment management may also direct responsible investing activities. Case studies like DBS and CapitaLand show the possibility of generating corporate development while obtaining sustainable results. To create a sustainable and resilient future, a holistic strategy that incorporates economic, social, and environmental factors is required.

References

1. Gillan SL, Koch A, Starks LT (2021) Firms and social responsibility: A review of ESG and CSR research in corporate finance. Journal of Corporate Finance 66:101889. https://doi.org/https://doi.org/10.1016/J.JCORPFIN.2021.101889
2. Li TT, Wang K, Sueyoshi T, Wang DD (2021) ESG: Research Progress and Future Prospects. Sustainability 2021, Vol 13, Page 11663 13:11663. https://doi.org/10.3390/SU132111663
3. Halbritter G, Dorfleitner G (2015) The wages of social responsibility — where are they? A critical review of ESG investing. Review of Financial Economics 26:25–35. https://doi.org/https://doi.org/10.1016/J.RFE.2015.03.004
4. Helmold M (2022) Marketing Management as Part of the Corporate Strategy. Management for Professionals Part F376:13–34. https://doi.org/https://doi.org/10.1007/978-3-031-10097-0_2/COVER
5. Sudusinghe JI, Seuring S (2020) Social Sustainability Empowering the Economic Sustainability in the Global Apparel Supply Chain. Sustainability 2020, Vol 12, Page 2595 12:2595. https://doi.org/10.3390/SU12072595
6. Munir N, Hasnain M, Roessner U, Abideen Z (2021) Strategies in improving plant salinity resistance and use of salinity resistant plants for economic sustainability. https://doi.org/10.1080/10643389.2021.1877033 52:2150–2196. https://doi.org/10.1080/10643389.2021.1877033
7. Ma Y, Liu Y (2019) Turning food waste to energy and resources towards a great environmental and economic sustainability: An innovative integrated biological approach. Biotechnol Adv 37:107414. https://doi.org/https://doi.org/10.1016/J.BIOTECHADV.2019.06.013
8. Yuen KF, Ong KW, Zhou Y, Wang X (2023) Social media engagement of stakeholders in the oil and gas sector: Social presence, triple bottom line and source credibility theory. J Clean Prod 382:135375. https://doi.org/https://doi.org/10.1016/J.JCLEPRO.2022.135375
9. Laosirihongthong T, Samaranayake P, Nagalingam SV, Adebanjo D (2019) Prioritization of sustainable supply chain practices with triple bottom line and organizational theories: industry and academic perspectives. https://doi.org/10.1080/09537287.2019.1701233 31:1207–1221. https://doi.org/10.1080/09537287.2019.1701233
10. Sen B, Yadav SK, Kumar G, et al (2023) Performance of eco-benign lubricating/cooling mediums in machining of superalloys: A comprehensive review from the perspective of Triple Bottom Line theory. Sustainable Materials and Technologies 35:e00578. https://doi.org/https://doi.org/10.1016/J.SUSMAT.2023.E00578
11. Braccini AM, Margherita EG (2018) Exploring Organizational Sustainability of Industry 4.0 under the Triple Bottom Line: The Case of a Manufacturing Company. Sustainability 2019, Vol 11, Page 36 11:36. https://doi.org/10.3390/SU11010036
12. The American Public Health Association's Joint Call to Action for Healthy Communities and Its Role in Planning, Designing, and Constructing Sustainable Infrastructure: Panel Discussion for the International Conference on Sustainable Infrastructure (ICSI) 2021
13. Adeniran AI (2020) The Improbable and the Sustainable Remedies. Migration Crises in 21st Century Africa 111–133. https://doi.org/10.1007/978-981-15-6398-0_5
14. Errera MR, Dias TA da C, Maya DMY, Lora EES (2023) Global bioenergy potentials projections for 2050. Biomass Bioenergy 170:106721. https://doi.org/10.1016/J.BIOMBIOE.2023.106721
15. Han X, Lu L, Zheng Y, et al (2019) A review on the key issues of the lithium ion battery degradation among the whole life cycle. eTransportation 1:100005. https://doi.org/10.1016/J.ETRAN.2019.100005
16. Li J, Lim K, Yang H, et al (2020) AI Applications through the Whole Life Cycle of Material Discovery. Matter 3:393–432. https://doi.org/https://doi.org/10.1016/J.MATT.2020.06.011
17. Frederiksen CS, Pedersen DB, Nielsen MEJ, Idowu SO (2021) Corporate Social Indices: Refining the Global Reporting Initiative. CSR, Sustainability, Ethics and Governance 271–291. https://doi.org/10.1007/978-3-030-62501-6_13/COVER

18. Vallet-Bellmunt T, Fuertes-Fuertes I, Flor ML (2023) Reporting Sustainable Development Goal 12 in the Spanish food retail industry. An analysis based on Global Reporting Initiative performance indicators. Corp Soc Responsib Environ Manag 30:695–707. https://doi.org/https://doi.org/10.1002/CSR.2382
19. Gulluscio C, Puntillo P, Luciani V, Huisingh D (2020) Climate Change Accounting and Reporting: A Systematic Literature Review. Sustainability 2020, Vol 12, Page 5455 12:5455. https://doi.org/10.3390/SU12135455
20. Jebe R (2019) The Convergence of Financial and ESG Materiality: Taking Sustainability Mainstream. American Business Law Journal 56:645–702. https://doi.org/https://doi.org/10.1111/ABLJ.12148
21. Hörisch J, Schaltegger S, Freeman RE (2020) Integrating stakeholder theory and sustainability accounting: A conceptual synthesis. J Clean Prod 275:124097. https://doi.org/https://doi.org/10.1016/J.JCLEPRO.2020.124097
22. Tettamanzi P, Venturini G, Murgolo M (2022) Sustainability and Financial Accounting: a Critical Review on the ESG Dynamics. Environmental Science and Pollution Research 29:16758–16761. https://doi.org/https://doi.org/10.1007/S11356-022-18596-2/METRICS
23. Li D, Huang M, Ren S, et al (2018) Environmental Legitimacy, Green Innovation, and Corporate Carbon Disclosure: Evidence from CDP China 100. Journal of Business Ethics 150:1089–1104. https://doi.org/https://doi.org/10.1007/S10551-016-3187-6/METRICS
24. Chen F, Hussain M, Khan JA, et al (2021) Voluntary disclosure of greenhouse gas emissions by cities under carbon disclosure project: A sustainable development approach. Sustainable Development 29:719–727. https://doi.org/https://doi.org/10.1002/SD.2169
25. Mildenberger M, Lachapelle E, Harrison K, Stadelmann-Steffen I (2022) Limited impacts of carbon tax rebate programmes on public support for carbon pricing. Nature Climate Change 2022 12:2 12:141–147. https://doi.org/10.1038/s41558-021-01268-3
26. Ramasubramanian, B., Rao, R. P., Chellappan, V., & Ramakrishna, S. (2022). Towards sustainable fuel cells and batteries with an AI perspective. Sustainability, 14(23), 16001.
27. Wu H, Sun Y, Su Y, et al (2022) Which Is the Best Supply Chain Policy: Carbon Tax, or a Low-Carbon Subsidy? Sustainability 2022, Vol 14, Page 6312 14:6312. https://doi.org/https://doi.org/10.3390/SU14106312
28. Ramadhani DP, Koo Y (2022) Comparative analysis of carbon border tax adjustment and domestic carbon tax under general equilibrium model: Focusing on the Indonesian economy. J Clean Prod 377:134288. https://doi.org/https://doi.org/10.1016/J.JCLEPRO.2022.134288
29. Luo R, Zhou L, Song Y, Fan T (2022) Evaluating the impact of carbon tax policy on manufacturing and remanufacturing decisions in a closed-loop supply chain. Int J Prod Econ 245:108408. https://doi.org/https://doi.org/10.1016/J.IJPE.2022.108408
30. Zhang Y, Qi L, Lin X, et al (2022) Synergistic effect of carbon ETS and carbon tax under China's peak emission target: A dynamic CGE analysis. Science of The Total Environment 825:154076. https://doi.org/https://doi.org/10.1016/J.SCITOTENV.2022.154076
31. Jayasiri NK, Kumarasinghe S, Pandey R (2023) 12 years of integrated reporting: A review of research. Accounting & Finance 63:2187–2243. https://doi.org/https://doi.org/10.1111/ACFI.12958
32. Velte P (2022) Does sustainable corporate governance have an impact on materiality disclosure quality in integrated reporting? International evidence. Sustainable Development 30:1655–1670. https://doi.org/https://doi.org/10.1002/SD.2333
33. Wu Y, Zhou S (2022) Do firms practicing integrated reporting engage in less myopic behavior? International evidence on opportunistic earnings management. Corporate Governance: An International Review 30:290–310. https://doi.org/https://doi.org/10.1111/CORG.12401
34. Qaderi SA, Ghaleb BAA, Hashed AA, et al (2022) Board Characteristics and Integrated Reporting Strategy: Does Sustainability Committee Matter? Sustainability 2022, Vol 14, Page 6092 14:6092. https://doi.org/https://doi.org/10.3390/SU14106092
35. Park S, Graham M, Foster EA (2022) Improving Local Government Resilience: Highlighting the Role of Internal Resources in Crisis Management. Sustainability 2022, Vol 14, Page 3214 14:3214. https://doi.org/10.3390/SU14063214

36. Levstek A, Pucihar A, Hovelja T (2022) Towards an Adaptive Strategic IT Governance Model for SMEs. Journal of Theoretical and Applied Electronic Commerce Research 2022, Vol 17, Pages 230–252 17:230–252. https://doi.org/10.3390/JTAER17010012
37. PricewaterhouseCoopers LLP - Company Profile and News - Bloomberg Markets. https://www.bloomberg.com/profile/company/10182Z:US#xj4y7vzkg. Accessed 22 Jul 2023
38. Aalberts RJ, Poon PS (1996) The New Prudent Investor Rule and the Modern Portfolio Theory: A New Direction for Fiduciaries. American Business Law Journal 34:
39. Ravikoff RB, Curzan MP (1980) Social Responsibility in Investment Policy and the Prudent Man Rule. Calif Law Rev 68:
40. Velte P (2023) Which institutional investors drive corporate sustainability? A systematic literature review. Bus Strategy Environ 32:42–71. https://doi.org/https://doi.org/10.1002/BSE.3117
41. Khan PA, Johl SK, Akhtar S, et al (2022) Open Innovation of Institutional Investors and Higher Education System in Creating Open Approach for SDG-4 Quality Education: A Conceptual Review. Journal of Open Innovation: Technology, Market, and Complexity 2022, Vol 8, Page 49 8:49. https://doi.org/10.3390/JOITMC8010049
42. Gary SN (2006) Charities, Endowments, and Donor Intent: The Uniform Prudent Management of Institutional Funds Act. Georgia Law Review 41:
43. Galle B (2017) Why Do Foundations Follow the Law? Evidence from Adoption of the Uniform Prudent Management of Institutional Funds Act. Journal of Policy Analysis and Management 36:532–556. https://doi.org/https://doi.org/10.1002/PAM.21987
44. Dathe T, Dathe R, Dathe I, Helmold M (2022) Corporate Social Responsibility (CSR) Versus Environmental Social Governance (ESG). Management for Professionals Part F361:117–141. https://doi.org/https://doi.org/10.1007/978-3-030-92357-0_9/COVER
45. Sandberg J (2011) Socially Responsible Investment and Fiduciary Duty: Putting the Freshfields Report into Perspective. Journal of Business Ethics 101:143–162. https://doi.org/https://doi.org/10.1007/S10551-010-0714-8/METRICS
46. Sopykhanova A, Maytanov A, Kiseleva A, Zhamiyeva R (2023) Problems of Legal Regulation and State Policy Measures Related to Nature Management in the Framework of Achieving the SDGs: Examples from Russia and Kazakhstan. Sustainability 2023, Vol 15, Page 1042 15:1042. https://doi.org/10.3390/SU15021042
47. Mekhantseva KF, Kuznetsov NG, Roschina LN, Rodionova ND (2023) ESG Management and Its Accounting and Analytical Reflection in the Practice of Sustainable Development in Russia at Different Territorial Levels. Environmental Footprints and Eco-Design of Products and Processes 563–573. https://doi.org/10.1007/978-3-031-28457-1_57/COVER
48. Orazalin N, Mahmood M, Narbaev T (2019) The impact of sustainability performance indicators on financial stability: evidence from the Russian oil and gas industry. Environmental Science and Pollution Research 26:8157–8168. https://doi.org/https://doi.org/10.1007/S11356-019-04325-9/METRICS
49. Kouloukoui D, Sant'Anna ÂMO, da Silva Gomes SM, et al (2019) Factors influencing the level of environmental disclosures in sustainability reports: Case of climate risk disclosure by Brazilian companies. Corp Soc Responsib Environ Manag 26:791–804. https://doi.org/10.1002/CSR.1721
50. Ibm IBM 2022 ESG Report and Addendum
51. Apple's 2022 ESG Report Environmental Social Governance Report
52. 2022 ESG Report: A year of impact | McKinsey. https://www.mckinsey.com/about-us/social-responsibility/esg-report-overview. Accessed 22 Jul 2023
53. 55. Tong SJW, Gan BQ, Tan KS (2022) Community structure of deep-sea benthic metazoan meiofauna in the polymetallic nodule fields in the eastern Clarion-Clipperton Fracture Zone, Pacific Ocean. Deep Sea Research Part I: Oceanographic Research Papers 188:103847. https://doi.org/https://doi.org/10.1016/J.DSR.2022.103847
54. Loh M, Ting J, Tong L, Deputy Y MEDIA RELEASE Unaudited Results of Keppel REIT for the Second Half and Full Year
55. Cowen T (2023) Singapore facts of the day, shareholder state edition

56. De La Orden R, De Calonje I Sustainability-Linked Finance-Mobilizing Capital for Sustainability in Emerging Markets FIGURE 1 Sustainable Debt Annual Issuance, 2013–2021 ($ billion)

Appendix

Biodiversity: The variety of life on earth, including ecosystems, species, and genetic diversity.

Biomimicry: The design and production of materials and systems inspired by nature and its processes.

Blue economy: An economic system that prioritizes the sustainable use of ocean resources and ecosystems.

Carbon footprint: The measure of the amount of greenhouse gases, primarily carbon dioxide, released into the atmosphere as a result of human activities.

Carbon offsetting: The practice of reducing greenhouse gas emissions by investing in projects that reduce or remove carbon from the atmosphere, such as reforestation or renewable energy.

Carbon–neutral: The state of producing no net carbon emissions by balancing carbon emissions with carbon removal or offsetting activities.

Circular bioeconomy: The use of renewable biological resources to create sustainable products and materials in a circular economy.

Circular design: The design of products or systems that prioritize resource efficiency, durability, and recyclability.

Circular economy: A system of production and consumption that seeks to maximize the use of resources and minimize waste.

Circular supply chain: A supply chain that operates in a circular economy, using resources efficiently and minimizing waste.

Climate adaptation: The process of preparing for and adapting to the impacts of climate change, such as sea level rise, increased frequency of extreme weather events, and changing precipitation patterns.

S. Ramakrishna and B. Ramasubramanian, *Handbook of Materials Circular Economy*,
https://doi.org/10.1007/978-981-97-0589-4

Climate mitigation: The process of reducing greenhouse gas emissions to prevent or minimize the impacts of climate change.

Closed-loop system: A system where waste is recycled and reused to create new products or materials, minimizing waste and reducing the need for new resources.

Corporate social responsibility: The responsibility of corporations to operate in an ethical and sustainable manner that benefits society and the environment.

Cradle-to-cradle: A design philosophy that aims to create products that can be fully recycled or biodegraded at the end of their useful life.

Decarbonization: The reduction of carbon emissions to combat climate change.

Downcycling: The process of converting waste materials into new products of lower value or quality.

Ecolabeling: The certification of products or services that meet certain environmental and sustainability standards.

Ecological economics: An approach to economics that integrates environmental and social factors into economic decision-making.

Ecological footprint: The measure of the impact of human activities on the natural environment in terms of the amount of land and water required to sustain them.

Ecosystem services: The benefits that humans derive from natural ecosystems, such as pollination, water filtration, and carbon storage.

Ecotourism: Tourism that prioritizes environmental sustainability, conservation, and education.

Energy efficiency: The efficient use of energy to reduce waste and increase productivity.

Environmental education: The education and awareness-raising of environmental issues, sustainability, and conservation.

Environmental justice: The fair and equitable distribution of environmental benefits and risks to all members of society.

Environmental stewardship: The responsible management of natural resources and ecosystems.

Ethical consumerism:

The practice of making purchasing decisions based on ethical and sustainable considerations.

Extended producer responsibility: The principle that manufacturers are responsible for the environmental impact of their products throughout their entire life cycle.

Food security: The availability and access to sufficient, safe, and nutritious food for all people.

Green chemistry: The design and production of chemicals that are environmentally sustainable and safe.

Green design: The design of products or systems that prioritize environmental sustainability and resource efficiency.

Green energy: Energy produced from renewable and sustainable sources, such as solar, wind, and hydropower.

Green infrastructure: Natural and engineered systems that provide environmental, economic, and social benefits, such as parks, green roofs, and wetlands.

Green jobs: Jobs that contribute to the development and implementation of environmentally sustainable practices and technologies.

Greenhouse gas emissions: The release of gases, such as carbon dioxide and methane, into the atmosphere that contribute to climate change.

Greenwashing: The practice of making false or misleading claims about the environmental sustainability of a product or service.

Life cycle assessment: A method used to evaluate the environmental impact of a product or process throughout its entire life cycle.

Life cycle thinking: A systems approach to understanding the environmental impact of a product or process throughout its entire life cycle.

Natural capital: The natural resources and ecosystem services that provide economic and environmental benefits.

Natural resource management: The sustainable management of natural resources, such as forests, fisheries, and water sources.

Permaculture: The design of sustainable and self-sufficient ecosystems based on natural patterns and processes.

Regenerative design: The design of products, systems, and processes that work to restore and improve natural ecosystems.

Renewable energy certificates: Certificates that represent the environmental attributes of renewable energy generation, used to support the growth of renewable energy.

Renewable energy: Energy derived from naturally replenishing sources, such as solar, wind, and hydropower.

Resilience: The ability of a system to adapt and recover from disturbances or shocks, such as natural disasters or economic crises.

Resource efficiency: The efficient use of resources to reduce waste and increase productivity.

Social entrepreneurship: The use of entrepreneurial principles and practices to create social and environmental benefits.

Sustainability: The ability to meet the needs of the present without compromising the ability of future generations to meet their own needs.

Sustainable agriculture: Agriculture that prioritizes environmental sustainability, social responsibility, and ethical practices.

Sustainable architecture: The design of buildings and structures that prioritize environmental sustainability, resource efficiency, and human well-being.

Sustainable business: Business practices that prioritize environmental sustainability, social responsibility, and ethical practices.

Sustainable cities: Cities designed and managed to prioritize environmental sustainability, social responsibility, and efficient use of resources.

Sustainable development goals: The 17 goals adopted by the United Nations to achieve sustainable development by 2030, covering social, economic, and environmental factors.

Sustainable development: Development that meets the needs of the present without compromising the ability of future generations to meet their own needs.

Sustainable fashion: The design, production, and consumption of fashion that prioritizes environmental sustainability, social responsibility, and ethical practices.

Sustainable forestry: The sustainable management of forests and forest resources, balancing environmental, social, and economic factors.

Sustainable materials: Materials that are environmentally friendly, durable, and recyclable or biodegradable.

Sustainable seafood: The sustainable production and consumption of seafood, based on environmental, social, and economic factors.

Sustainable sourcing: The sourcing of materials and resources that prioritize environmental sustainability, social responsibility, and ethical practices.

Sustainable tourism: Tourism that prioritizes environmental sustainability, conservation, and education.

Sustainable transportation: Transportation that prioritizes environmental sustainability, social responsibility, and efficient use of resources.

Triple bottom line: An accounting framework that evaluates the economic, social, and environmental performance of a company or organization.

Upcycling: The process of converting waste materials into new products of higher value or quality.

Water conservation: The efficient use and management of water resources to reduce waste and increase sustainability.

Zero waste: The goal of producing no waste by reducing, reusing, and recycling materials.

Pay-as-you-throw (PAYT)—A waste management system where households and businesses are charged based on the amount of waste they generate, typically aimed at reducing waste and increasing recycling.

Shared value—A business concept that emphasizes creating value for both the company and society, typically achieved through sustainability initiatives that benefit the environment and local communities.

Life cycle assessment (LCA)—A tool used to evaluate the environmental impact of a product throughout its entire life cycle, from raw material extraction to disposal.

Material flow analysis (MFA)—A method used to analyze the flow of materials within a system or economy, typically used to identify opportunities to improve resource efficiency and circularity.

Standards to refer for LCA and sustainability reporting

ISO 14040: Environmental management—Life cycle assessment—Principles and framework

ISO 14044: Environmental management—Life cycle assessment—Requirements and guidelines

ISO 14025: Environmental labels and declarations—Type III environmental declarations—Principles and procedures

ISO 14046: Water footprint—Principles, requirements and guidelines

ISO 14064-1: Greenhouse gases—Part 1: Specification with guidance at the organization level for quantification and reporting of greenhouse gas emissions and removals

ISO 14064-2: Greenhouse gases—Part 2: Specification with guidance at the project level for quantification, monitoring and reporting of greenhouse gas emission reductions or removal enhancements

ISO 14064-3: Greenhouse gases—Part 3: Specification with guidance for the validation and verification of greenhouse gas assertions

ISO 14067: Carbon footprint of products—Requirements and guidelines for quantification and communication

ISO 14080: Framework and principles for assessing and reporting on sustainable value creation

PAS 2050: Specification for the assessment of the life cycle greenhouse gas emissions of goods and services

PAS 2070: Specification for the assessment of greenhouse gas emissions of a city by a corporate or organizational value chain approach

PAS 2080: A Framework for Embedding the Principles of Sustainable Development in Infrastructure Projects

EN 15804: Sustainability of construction works—Environmental product declarations—Core rules for the product category of construction products

ASTM E2129-13: Standard Practice for Data Collection for Sustainability Assessment of Building Products

ASTM E3012-16: Standard Guide for Characterizing Environmental Aspects of Manufacturing Processes

WRI/WBCSD GHG Protocol Corporate Accounting and Reporting Standard

WRI/WBCSD GHG Protocol Product Life Cycle Accounting and Reporting Standard

The Social and Environmental Responsibility Audit (SECR) Standards (SECR-U and SECR-S)

ISEAL Code of Good Practice for Assessing the Impacts of Social and Environmental Standards Systems

EU Product Environmental Footprint (PEF) and Organisation Environmental Footprint (OEF) Guide.

PAS 2060: Specification for the demonstration of carbon neutrality

PAS 2395: Sustainability criteria for bioenergy

ASTM E1996-11: Standard Specification for Performance of Exterior Windows, Curtain Walls, Doors, and Impact Protective Systems Impacted by Windborne Debris in Hurricanes

ISO 14015: Environmental management—Environmental assessment of sites and organizations (EASO)

ISO/IEC 17020: Conformity assessment—Requirements for the operation of various types of bodies performing inspection

ISO/IEC 17025: General requirements for the competence of testing and calibration laboratories

ISO 50001: Energy management systems—Requirements with guidance for use

EN 16258: Methodology for calculation and declaration of energy consumption and GHG emissions of transport services (freight and passengers)

EN 15804: Sustainability of construction works—Environmental product declarations—Core rules for the product category of construction products.

PAS 2030: Improving the energy efficiency of existing buildings—Specification for installation process, process management and service provision.

ASTM D6866: Standard Test Methods for Determining the Biobased Content of Solid, Liquid, and Gaseous Samples Using Radiocarbon Analysis

ASTM E2635: Standard Practice for Water Conservation in Buildings Through In-Situ Water Reclamation

ASHRAE Standard 189.1: Standard for the Design of High-Performance Green Buildings Except Low-Rise Residential Buildings

Global Reporting Initiative (GRI) Standards: The GRI Protocol for Greenhouse Gas (GHG) Emissions is a collection of recommendations designed to assist companies in measuring and reporting their greenhouse gas emissions. This protocol is part of the GRI Guidelines, which give a framework for reporting on sustainability. According to the GRI Protocol for GHG Emissions, companies must report on three types of emission levels, known as Scope 1, Scope 2, and Scope 3 emissions:

Emissions of Scope 1: These are direct emissions from sources owned or controlled by the reporting entity. Emissions from the burning of fuels in boilers, furnaces, cars, and other equipment owned by the organization are examples of Scope 1 emissions.

Scope 2 Emissions: These are indirect emissions resulting from the purchase of energy, heat, or steam by the reporting entity. Scope 2 emissions are related with the generation of electricity by power plants and other sources that supply the organization with electricity.

Scope 3 Emissions: All additional indirect emissions not included in Scope 2 emissions. Emissions from the extraction and manufacturing of bought materials and fuels, transportation of goods and waste, and disposal of waste created by the organization are examples of Scope 3 emissions.

Many companies including Unilever, Nestle, Nike, Coca cola, and Microsoft does report their sustainability using GRI protocols. The following are some of the important points from *Unilever's 2020 GRI sustainability report*:

Since 2008, its industrial activities have reduced greenhouse gas emissions by 61%.

Since 2008, there has been a 36% reduction in water abstraction per tonne of output.

77% of its agricultural raw materials are sourced responsibly.

The establishment of a €1 billion Climate and Nature Fund to fund reforestation, wildlife conservation, and other nature-based climate change solutions.

https://www.unilever.com/planet-and-society/sustainability-reporting-centre/sustainability-performance-data/

GRI Protocol for Water: This protocol provides guidance on measuring and reporting water use and discharge, and sets out requirements for reporting on water consumption, withdrawal, and discharge. This protocol includes reporting requirements for water management, such as setting targets and implementing water efficiency measures, as well as for monitoring and disclosing water-related impacts and risks. https://www.globalreporting.org/how-to-use-the-gri-standards/gri-standards-english-language/

GRI Protocol for Waste: This protocol provides guidance on measuring and reporting waste generation, disposal, and recycling, and sets out requirements for reporting on the type and quantity of waste generated. This protocol includes reporting requirements for waste management, such as setting targets and implementing waste reduction measures, as well as for monitoring and disclosing waste-related impacts and risks.

GRI Protocol for Labor Practices and Decent Work: This protocol provides guidance on reporting on labor practices and decent work, and sets out requirements for reporting on employment practices, working conditions, and human rights. This protocol includes reporting requirements for workforce demographics, health and safety, employee engagement, training and development, and supplier labor practices.

GRI Protocol for Human Rights: This protocol provides guidance on reporting on human rights and sets out requirements for reporting on the organization's policies and practices related to human rights. This protocol includes reporting requirements for assessing human rights risks and impacts, implementing due diligence measures, and engaging with stakeholders on human rights issues.

GRI Protocol for Society: This protocol provides guidance on reporting on the organization's impact on society, and sets out requirements for reporting on community engagement, philanthropy, and other social impact indicators. This protocol includes reporting requirements for stakeholder engagement, community investments, and social impact assessments.

GRI Protocol for Product Responsibility: This protocol provides guidance on reporting on the organization's product responsibility, and sets out requirements for reporting on product safety, labelling, and other product-related indicators. This protocol includes reporting requirements for product design, safety and quality assurance, labelling and disclosure, and responsible marketing and advertising.